CW00531047

Physicists of Ireland
Passion and Precision

Physicists of Ireland
Passion and Precision

Edited by

Mark McCartney
University of Ulster

and

Andrew Whitaker
Queen's University Belfast

Institute of Physics Publishing
Bristol and Philadelphia

British Library Cataloguing-in-Publication Data
A catalogue record for this book is available from the British Library.

ISBN 0 7503 0866 4

Library of Congress Cataloging-in-Publication Data are available

Commissioning Editor: James Revill
Production Editor: Simon Laurenson
Production Control: Sarah Plenty
Cover Design: Frédérique Swist
Marketing: Nicola Newey and Verity Cooke

Published by Institute of Physics Publishing, wholly owned by The Institute of Physics, London

Institute of Physics Publishing, Dirac House, Temple Back, Bristol BS1 6BE, UK

US Office: Institute of Physics Publishing, Suite 929, The Public Ledger Building, 150 South Independence Mall West, Philadelphia, PA 19106, USA

Typeset by Academic + Technical, Bristol
Printed in the UK by MPG Books Ltd, Bodmin, Cornwall

How should the world be luckier if this house,
Where *passion and precision* have been one
Time out of mind, became too ruinous
To breed the lidless eye that loves the sun?
And the sweet laughing eagle thoughts that grow
Where wings have memory of wings, and all
That comes of the best knit to the best? Although
Mean roof-trees were the sturdier for its fall,
How should their luck, run high enough to reach
The gifts that govern men, and after these
To gradual Time's last gift, a written speech
Wrought of high laughter, loveliness and ease?

W B Yeats, *Upon a House Shaken by the Land Agitation*

Contents

Preface

Ireland has produced and nurtured physicists from the time of Newton to the present day. A number of the physicists whose lives and works are recorded in this book are among the most important figures in the history of physics, for example Boyle, Stokes, Hamilton, Kelvin, Schrödinger and Bell. A whole host of others have left an indelible mark in the development of science, their names being well known to any student of physics or chemistry. These would include, for example, Andrews, Joly, Townsend, Synge and Walton, and there are many other representatives in this book. We have also included a few figures, who, though not widely remembered today, not only made substantial contributions to the development of physics, but also played an important part in the development of science and scientific institutions in Ireland, for example Petty, Molyneux and Haughton.

The editors are not chauvinistic enough to believe that there is, or should be, a specifically 'Celtic Physics' (though there well may be an Irish tradition in physics, a point touched on by several of our contributors). It does, though, seem a useful exercise to highlight the contribution made to physics by those with Irish connections. It may be useful, and even pleasant, for people in Ireland itself, to encourage pride in past achievement, and perhaps even a heightened interest in present endeavours, and it may also be useful for those outside Ireland, who perhaps are unaware of our contributions, and who may imagine that Kelvin, Stokes and the like are straightforwardly 'English' or 'British' physicists.

To qualify for inclusion, a contender obviously needed to have made a major contribution to physics; the 'territorial' qualification was either to be have strong Irish roots—thus Stokes, Larmor, Kelvin and Bell are included, even though their achievements were largely outside Ireland; or to have contributed in a major way to physics while in Ireland. In the latter category, we have included Schrödinger, Heitler and Lanczos; far from feeling guilty at including subjects so obviously non-Irish by origin (though Heitler took out Irish nationality), we feel that Ireland may take pride in providing a haven for such eminent scientists, thus enabling them to continue their excellent work.

Merely to have been born in Ireland was not enough for inclusion. Thus we had to exclude such fine scientists as Osborne Reynolds and Kathleen Lonsdale, who were born in Ireland, but, we felt, lacked strong enough Irish roots to be included. Sadly we also felt we could not include Harrie Massey, a major atomic physicist who spent a period at Queen's University Belfast; we felt, on balance, that he spent too little time in Ireland to justify inclusion, though it is true that he initiated David Bates, who, of course, *is* included, into research on atomic physics, and thus was primarily responsible for the extremely strong research group on atomic and molecular physics which continues at Queen's to this day. Lastly, and again sadly, we felt we could not include George Boole; his work at Cork would certainly satisfy the Irish connection, but unfortunately we felt we could not categorize his mathematical achievements as 'physics', important for physicists as they undoubtedly are. (More prosaically we would mention that we have not included subjects, however worthy, who are still happily alive.)

Although this book is a collection of articles on individual physicists and not a history of physics in Ireland, it does allow the reader to gain a feel for the development of Irish physics and the sense of community amongst Irish physicists of the same era. In the contributions we have indicated with an asterisk links between different physicists covered in the book. These are numerous, particularly in the Victorian era, a fact which encourages one to wonder whether there was an element of Irish physicists promoting and supporting one another. This idea has been discussed by Norman McMillan in his book *Prometheus's Fire*. The present book may provide further data in regard to this hypothesis, but we do not attempt to draw any conclusion.

Our authors have endeavoured to make their contributions accessible to those who are interested in science, but do not necessarily have an extensive knowledge of the subject. We feel they have made their contributions not only readable and interesting, but also scholarly. In particular we hope that sixth formers and undergraduates will read these contributions and be enthused by the strong tradition of Irish physics.

As editors we would like to thank wholeheartedly all our contributors who readily agreed to give their time to this project and reacted positively to all our requests and suggestions. This has allowed us to achieve generally a consistent format and style across the book. We would mention, though, one area in which we have not sought full consistency. Most of our subjects died some time ago, it might be said that they belong to history, and the tone in which they are discussed in this book is naturally impersonal. However, for those physicists who have died more recently, we have chosen contributors who were not only familiar with the work, but also close to the actual subject. Their contributions have naturally been more personal in nature, a fact we consider a strength rather than a weakness.

In many cases our contributors have also been kind enough to provide illustrations. We would like to thank them for this and all others (particularly Alistair Montgomery, Ken Bell, Martin Lamb, Jennifer McKee, Wendy Rutherford and Lynne Hamilton) who have helped us with illustrations. We would also like to thank the staff of the Science Library at Queen's University Belfast for helping check references. We would acknowledge all those who have given permission to publish illustrations of which they have the copyright. In some cases we have not been able to establish and acknowledge the owner of the copyright; where such cases are brought to our attention, we will make amends in future editions of this work.

Among our contributors, particular thanks are due to Norman McMillan, Roy Johnston, Petros Florides, James O'Hara, Tom O'Connor and Denis Weaire who made useful suggestions about potential subjects and potential contributors, and gave helpful general advice. We would also thank the Walton family who suggested asking Brian Cathcart to write on Ernest Walton.

We would acknowledge the excellent collection, *Creators of Mathematics: The Irish Connection*, which was edited by Ken Houston and published by University College Dublin Press in September 2000; we would acknowledge that this book stimulated the preparation of the present volume. Six subjects are common to both books, but, since we felt it would be profitable to obtain a diversity of views, in only one case did we commission an author to write on the same subject as in the earlier book. Incidentally Ken came to a different decision from us on the inclusion of Osborne Reynolds; readers disappointed in finding no article on him in this book will find one in his.

We would also like to recommend the interesting and beautifully illustrated collections edited by Charles Mollan, William Davis and Brendan Finucane, *Some People and Places in Irish Science and Technology* and *More People and Places in Irish Science and Technology*, both published by the Royal Irish Academy in 1985 and 1990 respectively. Many of our contributors acknowledge these sources individually. In addition several physicists who might have been included in our book are discussed in these volumes: Robert Ball in the first volume, Walter Hartley, Alexander Anderson, Thomas Preston, Arthur Conway, John and Patrick Nolan, and Thomas Nevin in the second. These two volumes have recently been combined and updated as *Irish Innovators in Science and Technology*.

On a more personal level, the editors would like to thank their families, Karen and Cameron McCartney, and Joan, John and Peter Whitaker for encouragement and patience during the lengthy task of editing this book.

To conclude, we will be happy if the present volume serves to increase the awareness of Irish physics and Irish physicists, both in

Ireland and elsewhere. Undoubtedly our subjects have combined a passion for their subject leading to contributions of individuality and flair, with the precision to ensure these contributions are reliable and lasting.

Mark McCartney
Andrew Whitaker
April 2002

Contributors

D Thorburn Burns is Professor of Analytical Chemistry at Queen's University Belfast. He is a Member of the Royal Irish Academy and a Fellow of the Royal Society of Edinburgh.

Brian Cathcart worked for Reuters and the London *Independent* newspaper before turning to full-time writing. He has written on the making of the British atomic bomb, and the Stephen Lawrence murder case. He is currently writing a book about Cockcroft, Walton and the Cavendish in 1928–32.

Stephen Coonan lectures on photography in the Dublin Institute of Technology.

Ray Flannery is Regents' Professor of Physics at Georgia Institute of Technology, USA. He is an Honorary Member of the Royal Irish Academy.

Raymond Flood is Senior Tutor at Kellogg College, Oxford, and University Lecturer in Computing Studies and Mathematics.

Petros S Florides is Associate Professor of Mathematics at Trinity College Dublin.

Barbara Gellai has been Senior Research Fellow in the Central Research Institute for Physics, Budapest, and Visiting Associate Professor of Physics at North Carolina State University. She is Associate Editor of the Lanczos Collected Papers, and an Independent Mathematical Consultant at Raleigh, USA.

David Glass is a Lecturer in Computing at the University of Ulster.

Roy Johnston has worked in high-energy physics and techno-economic modelling. He has managed an applied scientific consultancy group, and is currently working with a software house on knowledge-based architecture.

Mark McCartney is a Lecturer in Mathematics at the University of Ulster.

Niall E McKeith is Technical Officer at the National University of Ireland, Maynooth. He is also Curator of the National Science Museum at Maynooth, which includes the Callan collection.

Norman D McMillan is Senior Lecturer in Physics and Course Director in Photonics at the Institute of Technology, Carlow.

John J O'Connor is a Lecturer in Pure Mathematics at St Andrew's University.

Thomas O'Connor is retired from his position as Statutory Lecturer in Experimental Physics at National University of Ireland, Galway.

James G O'Hara has qualifications in electrical engineering and in the history of science and technology. He is Research Editor for the Leibniz Edition, based at the State Library of Lower Saxony in Hannover.

Edmund F Robertson is Professor of Mathematics at St Andrew's University.

Siddhartha Sen is Associate Professor of Mathematics at Trinity College Dublin. He is a member of the Royal Irish Academy.

Bob Strunz is a Teaching Technologist at the University of Limerick, and a Board Member of the Birr Educational Trust, which develops educational activities at the Birr estate, home of the Rosse family and the giant telescope.

Patrick Wayman was Senior Professor in the Astronomy Section of the School of Cosmic Physics of the Dublin Institute of Advanced Studies, Director of the Dunsink Observatory, and Honorary Andrews Professor at Trinity College Dublin. He had been General Secretary of the International Astronomical Union. He died in 1998.

Denis Weaire is Erasmus Smith's Professor of Natural and Experimental Philosophy in the Physics Department of Trinity College Dublin. He is a Fellow of the Royal Society, and a Member of the Royal Irish Academy and the Academia Europaea.

Andrew Whitaker is Professor of Physics at Queen's University Belfast.

Iwan Williams is Professor and Dean of the Faculty of Engineering and Mathematical Sciences, Queen Mary, University of London.

Alastair Wood is Wescan Professor of Applied Mathematics in Dublin City University.

NOTE

An asterisk (*) by the name of a scientist mentioned in an article is an indication that that scientist is the subject of a separate article in the book.

William Petty

1623–1687

Norman D McMillan

William Petty was a man of many parts, and this article focuses specifically on his role in Ireland, and more specifically his contributions to Irish institutions. He was an adventurer who came to Ireland as part of the Cromwellian army. He was born in Romsey, Hampshire, the eldest son of Anthony Petty, a cloth worker who perhaps was also a small landowner. He found amusement as a boy watching various local craftsmen, and then, aged 13, went to sea as a cabin boy for a short period and studied navigation. He attended a Jesuit college in Caen for a year after an accident at sea, returning to England to join the navy. Shortly thereafter he went to the Netherlands to study medicine. He was introduced to natural philosophy by Hobbes in Paris and joined a famous circle of savants that included Mersenne. He returned to England in 1646, and by 1649 was in Oxford University at the time the Royal Society of London was formed; it soon began holding meetings in his lodgings. He received his doctorate in physic in 1650.

His career from this point was meteoric. He was appointed Professor of Anatomy in Brasenose College, Oxford, and then Professor of Music in Gresham College, London. In 1652 he vacated both positions to become Physician-General for Cromwell's army in Ireland. In 1655 Petty volunteered to carry out the survey of Ireland in 13 months, provided he was able to engage 1000 men for the task. The important Down Survey was indeed completed in March 1656 on schedule. Later he is believed to have been involved in the first census of Ireland, held around 1659, as results of this census were found in his papers. His map of Ireland was then the most detailed ever produced; it was not printed until 1685. Other more detailed maps remained unpublished.

The money he raised in his work on the Down Survey and other speculation allowed him to buy much of the best land in Ireland cheaply; much of this land had been forfeited and mortgaged. He thereby acquired a vast landholding, but also lay himself open to hostility and litigation for the rest of his life. Despite being a personal friend of the Cromwell family, he acquiesced in the Restoration, and indeed was knighted by Charles II in 1661 for his role as a founding member of the Royal Society; perhaps

William Petty.

this was a sop to this Baconian society. He was, surprisingly, on good terms with the Catholic monarch, James II, who made his widow Baroness Shelburn in 1688.

In 1650 Petty conducted anatomical experiments in Ireland together with Robert Boyle*. The latter was not overly impressed with the country; he held that Ireland was a 'barbarous country, where chemical spirits were so misunderstood and chemical instruments so unprocurable, that it is hard to have any Hermetic thoughts of it'.

Petty was in Ireland in 1683 and gave active support to William Molyneux* in the foundation of the Dublin Philosophical Society as a sister to the Royal Society of London, which he himself had helped to found. He added prestige to the new society as its President from 1684. He helped frame the rules of the society, and drew up a new series of advertisements, containing some proposals for modelling 'our future progress'. He emphasized the cooperative nature of the work needed, and urged that time at meetings should be devoted to the carrying out of experiments. (His advice was not followed; meetings were usually restricted to reports of work done by others.) He also placed a heavy emphasis on the quantitative side of scientific researches. There was an entrance fee to the Society of 8s 6d, and an annual charge of 52s. Despite these very considerable sums, by 1685 the Society had grown in number to more than 30 'very sufficient men'. Petty was also instrumental in getting the Society to correspond with both the London and Oxford groups, and to build up other international relationships.

Petty's main contributions to Irish institutions arose from a small work *The Advice of W P to Samuel Hartlib,* which he published as a young man in the form of a pamphlet of just 26 pages. In this he set out his ideas for a Gymnasium Mechanicum for the Advancement of Mechanical Arts and Manufacture. This work suggested means of implementation of Bacon's ideas for just such an institution; it advanced an appeal for the public endowment of research and invention. Petty also proposed an Ergasula Literaria—Universal Workshop (Vernacular School). Petty's ideas were widely circulated and indeed almost became a prescription for a reform programme pursued by the Puritan enthusiasts of the New Learning, who came specifically to Dublin during the Commonwealth for just such projects of social reconstruction.

What actually materialized in Ireland was not one large institution in Dublin, but rather a diverse variety of institutions that, in total, may be felt to correspond quite closely to the ideas of Petty. The author has argued that in Ireland from the eighteenth century, the Protestants in particular pursued this project in a piecemeal fashion. Various projects developed that in time saw almost the complete edifice of Petty's 'grand scheme' constructed in Dublin.

His Academy for Pure Science came into being with the foundation of the Royal Irish Academy in 1785; Academic Infirmaries developed in many places around the city, though none perhaps was related specifically to the ideas in the Petty plan; and Botanic Gardens were established, initially in Trinity College but more permanently in Glasnevin in 1790. The Royal Dublin Society, which was established in 1731, was a cradle for these projects; the Society itself was in many regards a close fit to Petty's conceptions set out in his *Advice.* It established a Drawing School in 1748 and the National Gallery in 1858, and made many other developments that can be traced back to original ideas of Petty. The Veterinary Establishment was established initially in 1800 and subsequently the Royal Veterinary College was established in 1895. Finally, Petty's insistence on the need for and importance of astronomical facilities was answered in Dublin with the establishment of Dunsink Observatory in 1787.

The Royal Dublin Society also pursued a vigorous technological educational programme, that grew in the late eighteenth century into a unique institute, internationally acclaimed, and with a full complement of professors in the sciences and engineering. After years of political agitation, principally by nationalists, this eventually led through various stages into the Royal College of Science, which later became a central part of University College Dublin.

The ideas of Petty on training of young children were also important and his ideas of hand and eye training, object lessons, vocational education, especially in agriculture for boys and hand crafts for girls, became in time the template for all future vocational education, including the Pestalozzian and Montessorian types. (For the latter, see *Prometheus's Fire*, listed in Further Reading.) The legacy of Petty is this area is definite and sustained, and it can be argued that this was his greatest contribution to Ireland.

Petty was hugely important for his development of the study of statistics as the servant of economics. It may be suggested that John Graunt founded the sciences of demography and statistics in 1662, with his book *Natural and Political Observations Mentioned in a Natural Index, and Made Upon the Bills of Mortality*. Graunt, however, was a friend of Petty and, it has been argued, that Petty was a co-author of this work; in any event it is certain that he made possible the publishing of this landmark book through his active encouragement of the project.

Between 1682 and 1687, Petty himself published ten brief essays on population, in which he identified a number of key economic factors. He was the first author to attempt to base economic policy upon statistical data. His work *Treatise on Taxes and Contributions*, which was published in 1662, discussed the economy of England and Ireland, providing an analysis of rents, and emphasizing the importance of labour as part of national wealth. This work provides the correct analysis of the origin of wealth, which is one of Petty's principal claims to fame.

He pointed out the benefits of the division of labour and the importance of foreign trade in his 1683 book *Another Essay in Political Arithmetick Concerning the Growth of the City of London*. His later work *Verbum Sapierti*, published in 1691, contains the first estimate of national income and *The Political Economy of Ireland*, published in the same year, is a discourse on political economy supported by economic geography. *Political Arithmetick* was published in 1690, and here Petty discusses comparative economics in a study of the wealth and economic policies of England and France.

A Treatise of Ireland was written in 1687 as private advice to James II, and was published only in 1899. Unfortunately it was ignored by this monarch who, in fact, brought to Dublin in that year an administration that attempted to destroy the Dublin Philosophical Society, and indeed everything else Petty and his fellow thinkers had sought to build.

Petty left a wealth of other material that remained unpublished after his death. His writings as an economic theorist were not surpassed before

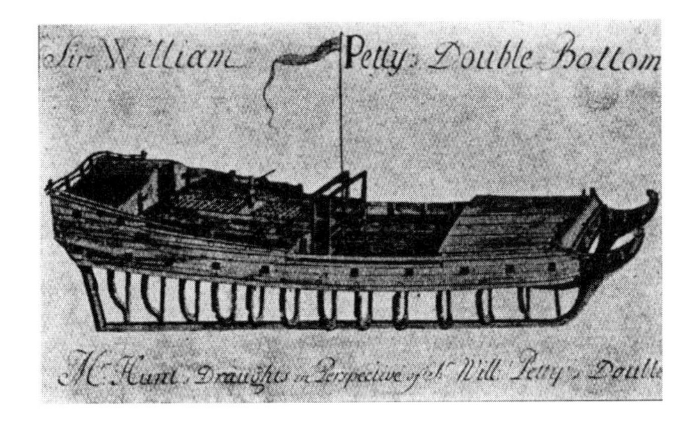

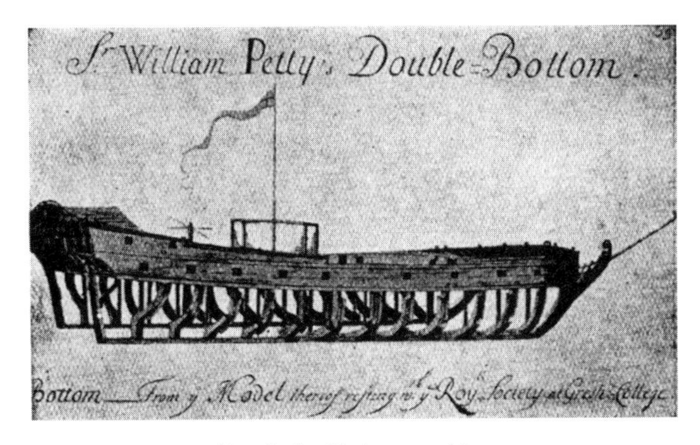

Petty's double-bottom ship.

1750, and he is widely acknowledged as the founder of the science of political economy.

We now move to Petty's major contributions to the physical sciences. Petty was a real example of what is known as virtuoso. He conducted investigations in anatomy, geodesy, and the design and testing of ships, he made studies of the practices of dyeing, and above all else he was a mathematician and statistician of real importance, and has become immortalized as the organizer of the greatest survey of his day.

Petty attempted to produce a powered ship fitted with an engine, and he invented 'a wheel to ride upon'. His most important contributions were, however, in marine and land transport engineering. From 1662 to 1664, he famously conducted experiments in Ireland on a double-keeled vessel. His work on marine engineering is seminal because of his use of a water tank and Lilliputian models which were towed in the tank to study their performance. The first two full-size ships built on the basis of this ship modelling were successful, but tragically the third sank in a storm with all hands aboard, and the fourth was also a complete failure. His concept was, however, fundamentally valid, the failures resulting from problems with the materials used, which could only be overcome in modern times. His dream was finally realized in our own day with the development of the catamaran.

Petty also designed a new kind of land carriage in collaboration with Sir William Spragge. His work on land carriages again involved building scale models, and this led to some practical advantages. Subsequently, Joshua Walker in Oxford developed this approach further, and came up with improvements concerning the dishing of cartwheels.

Petty had an exceptionally wide range of further accomplishments. He urged the study of number, weight and measurement. His practical examples in this regard concerned the scale and strengths of structures, and calculations on the speed of travel of sound, odour and light. He made suggestions on elasticity, and proposed an atomic theory of matter in which, interestingly, atoms were tiny magnets of different sexes. He was probably the first in Dublin to take an interest in mineral waters, which was a passion followed up subsequently by Berkeley. Lastly he supervised the work of Dr Allen Mullen, who published an important pamphlet on his work on the anatomized (or dissected) elephant.

At the end of his life he read *The Principia* in 1687, and concluded that this was a truly great work. He said 'Poor Mr Newton, I have not met with one person that put an extraordinary value on this book...I would give 500 pounds to have been the author of it; and 200 pounds that Charles [his son] understand it'. For someone so notoriously astute with money, this was the highest praise!

Further Reading

Article by Frank N Egerton in *Dictionary of Scientific Biography*, C C Gillespie (ed) (Scribner, New York, 1970–80). This entry on Petty gives very detailed references on all aspects of his life and work, and is an excellent starting point for any interested researcher to find the important references on Petty.

The Common Scientist in the Seventeenth Century: A Study of the Dublin Philosophical Society 1683–1708, K Theodore Hoppen (Routledge and Kegan Paul, London, 1970) gives information on the Dublin Philosophical Society and Petty's researches in physical science and engineering together with their context.

The Economic Writings of Sir William Petty together with Observations upon the Bills of Mortality, more probably by Captain John Graunt, W Petty

(Cambridge University Press, 1899; facsimile edition, New York, 1963) is the best source for Petty's economic and statistical writings. See also *Contributions to the History of Statistics,* H Westergaard (London, 1932; facsimile edition, New York, 1968).
The Histories of the Survey of Ireland Commonly Called the Down Survey, A.D. 1655–6, T A Larcom (ed) (Irish Archaeological Society, Dublin, 1851; facsimile edition, New York 1967).
Sir William Petty F.R.S. (1623–1687), in *The Royal Society, Its Origins and Founders,* H Hartley (ed) (Royal Society, London, 1960).
Prometheus's Fire: A History of Scientific and Technological Education in Ireland, N D McMillan (Tyndall, Carlow, 2000), pp 74–137 describes Petty's achievements in an Irish context.

Robert Boyle

1627–1691

D Thorburn Burns

Robert Boyle was born on 25 January 1627 in Lismore Castle, County Waterford, Ireland. He was the fourteenth child, and seventh and last son of Richard Boyle, the First Earl of Cork, and his wife, Lady Catherine Fenton.

His life and times may conveniently be divided into seven periods: Irish childhood (1627–1635), School at Eton (1635–1638), Manor at Stalbridge (1636–1638), Grand Tour (1638–1644), Stalbridge (1645–1655), Oxford (1656–1668) and London (1668–1691).

Robert Boyle's autobiographical sketch, *An Account of Philatetus during his Minority*, covers the first four periods. It was included in the works collected by Birch in 1744 (see Further Reading.) From internal evidence, this account was written after January 1648, and before July 1649.

Robert, or Robyn to the close family, was raised, as was the tradition at that time in wealthy families, by a wet nurse (foster mother) and then taught by private tutors at home. He was sent at the age of eight, along with his brother Francis aged 12, to Eton; it is reported that he was a good student. Some holiday time was spent at the Manor of Stalbridge in Dorset, purchased by the Earl in 1636.

After leaving school and a brief stay in the House of Savoy in London, he and Francis were sent under the charge of Isaac Marcombes to undertake a Grand Tour in Europe. Marcombes had earlier dealt with the Grand Tour for Robert's brothers, Lewis and Rodger Boyle. Just before departure Francis was married to Elizabeth Killigrew, a maid of honour at the Court; King Charles I gave the bride away. Four days later he was separated from his bride and sent to the continent. According to the Earl, it was never too early to think about a good match! Only two of the Earl's children escaped, and that for the sole reason that the Earl had not completed the arrangements at the time of his death.

When Robert was a mere 14 years old, and out of the country at that, the Earl rode out to Hatfield and presented a gold and diamond ring to the charming Lady Ann, daughter of Lord Howard of Escrick, as a pledge that she would become the wife of his youngest son. In his will

The Honourable Robert Boyle.

the Earl left her a wealth of silver plate that would be hers if she married Robert. Upon return from the Grand Tour, Robert made no serious attempt to follow his father's lead with the Howards, probably because he had witnessed too many loveless alliances. The ring was returned; in due course Lady Ann married her cousin Charles Howard, later First Earl of Carlisle.

The years 1638–1644 were spent making the Grand Tour of Europe. They got as far south as Italy and spent time studying the works of Galileo. Two years were spent in Geneva, where much time was devoted to mathematics and various language studies. One summer evening in 1640, Robert awoke to find himself in the midst of a violent thunderstorm and began to wonder why he was not struck by lightning. He came to the conclusion that God must have reserved some special task for him. Thereafter he dedicated himself to the demonstration of God's majesty by unravelling the secrets of nature. He was a devoted Christian all the rest of his days.

After his return to England in mid-1644, Robert lived with his widowed sister, Katherine, Viscountess Ranelagh, before moving to

Stalbridge Manor, part of his inheritance from his father. Here he sat out the Civil War and early Commonwealth period, and started on his largely self-taught career of chemical and physical experimentation. He was, however, back and forth to London, and intimate with the 'Invisible College', a circle of friends and the embryo of the Royal Society.

Boyle was back in Ireland from 1652 to 1654 dealing with family estate business; in a letter to Frederick Clodius written from Ireland he was the first to use the term 'chemical analysis' in the way it has been used since by chemists. Boyle moved to Oxford in 1656, and he lived in a house, known as 'Deep Hall', on the west side of University College. In Oxford he worked privately, with paid assistants such as Robert Hooke, mainly to provide proofs of the corpuscularian, mechanical theory of nature. He also studied medicine. Life in Oxford was congenial, for many of the 'Invisibles' were resident in that City.

Robert Boyle's last move was to London in 1668 so as to be near the Royal Society of which he was a founder member. The move should have occurred earlier, but was delayed due to the plague of 1665. He lived with his sister in Pall Mall, next door but one to Nell Gwynne, until his death in 1691. The house and land included laboratory facilities.

Boyle's scientific studies fell mainly into the areas of physics and chemistry. Those in chemistry including analytical chemistry are mainly excluded from this article; detailed information and comment are available in the works of Boas and Thorburn Burns listed in Further Reading. His main contributions to physics were through his work on gases, heat and thermometry, light and colours, accurate density measurements and, to a lesser extent, on magnetism and electricity.

Boyle's first scientific publication, written in 1660 after six years' work in Oxford, was *New Experiments Physico-Mechanical, Touching the Spring of the Air, and its Effects, Made, for the most part, in a New Pneumatical Engine.* With this engine, or air pump, Boyle could demonstrate that air transmitted sound and had weight. In the book he also discussed lode-stones, or natural magnets, and described chemical experiments, such as the demonstration that a candle required air to continue burning.

This book helped to popularize the experimental approach to science; it was a scientific work that everybody could understand. However shortly after its publication a few criticisms emerged among the generous praise given by most scientists worldwide. One scholar in particular, Franciscus Linus, a Belgian physics professor at the University of Liège, offered an entirely different explanation of why mercury stays up in a barometer tube. Linus claimed that an invisible cord, or *funiculus,* drew the mercury up, and the rise had nothing to do with the external air.

Boyle decided that the best way to answer Linus would be by experiments rather than long letters in scientific journals. In these experiments Boyle discovered the inverse relationship between the

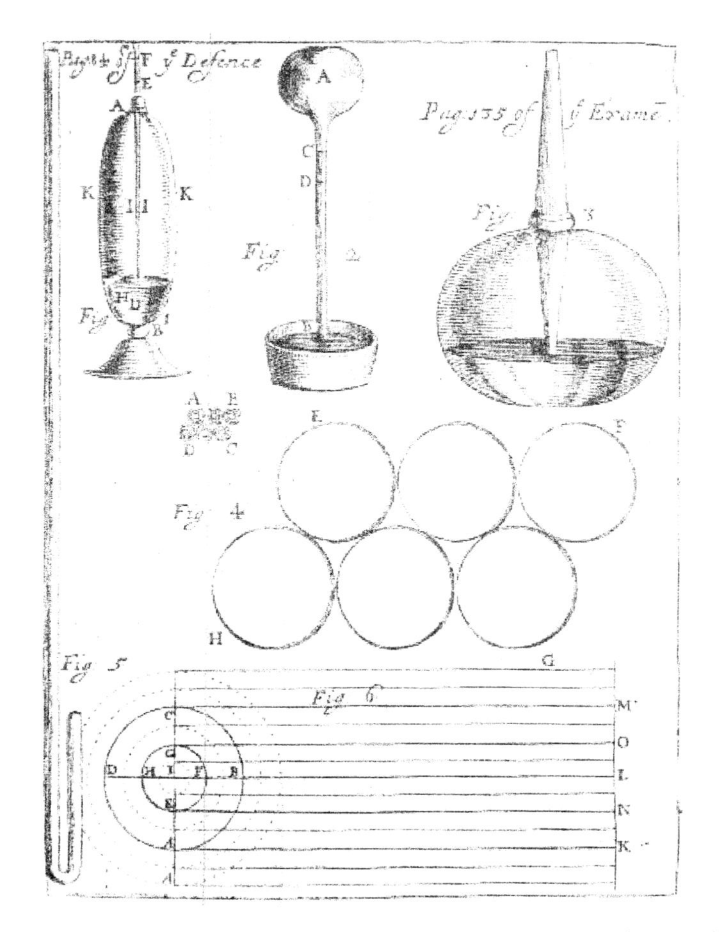

Plate given in Boyle's defence against Linus, taken from the Leers edition, 1669.

pressure and volume of a sample of gas. These experiments were published in an appendix annexed to a reprint to *New Experiments*... published in 1662 and called *A Defence of the Doctrine touching the Spring and Weight of the Air, Against the Objections of Franciscus Linus. Wherewith the Objector's Funicular Hypothesis is also Examin'd.*

The experiments were quite simple; air was trapped in a sealed U-tube, and the volume and head of pressure were read from paper scales glued to the tubes. More mercury was added, the apparatus was allowed to settle, and the volume of the compressed gas and the head of pressure were again read. The pressure head was added to the barometric pressure to give the pressure of the sealed-off gas space. Boyle then calculated the pressure expected on the basis of the hypothesis 'that the pressures and the expansions to be in reciprocal proportion'. That this was so is shown by the agreement of the data in columns D

(60)

diviſions in the ſhorter Tube, the ſeveral Obſervations that were thus ſucceſſively made, and as they were made ſet down, afforded us the the enſuing Table.

A Table of the Condenſation of the Air.

A	A	B	C	D	E
48	12	00	Added to $29\frac{1}{8}$ makes	$29\frac{2}{16}$	$29\frac{2}{16}$
46	$11\frac{1}{2}$	$01\frac{7}{16}$		$30\frac{9}{16}$	$30\frac{6}{16}$
44	11	$02\frac{13}{16}$		$31\frac{15}{16}$	$31\frac{12}{16}$
42	$10\frac{1}{2}$	$04\frac{6}{16}$		$33\frac{8}{16}$	$33\frac{1}{7}$
40	10	$06\frac{3}{16}$		$35\frac{5}{16}$	35 --
38	$9\frac{1}{2}$	$07\frac{14}{16}$		37 --	$36\frac{15}{19}$
36	9	$10\frac{2}{16}$		$39\frac{4}{16}$	$38\frac{7}{8}$
34	$8\frac{1}{2}$	$12\frac{8}{16}$		$41\frac{10}{16}$	$41\frac{2}{17}$
32	8	$15\frac{1}{16}$		$44\frac{3}{16}$	$43\frac{11}{16}$
30	$7\frac{1}{2}$	$17\frac{15}{16}$		$47\frac{1}{16}$	$46\frac{3}{5}$
28	7	$21\frac{3}{16}$		$50\frac{5}{16}$	50 --
26	$6\frac{1}{2}$	$25\frac{3}{16}$		$54\frac{5}{16}$	$53\frac{10}{13}$
24	6	$29\frac{11}{16}$		$58\frac{13}{16}$	$58\frac{2}{8}$
23	$5\frac{3}{4}$	$32\frac{3}{16}$		$61\frac{5}{16}$	$60\frac{18}{23}$
22	$5\frac{1}{2}$	$34\frac{15}{16}$		$64\frac{1}{16}$	$63\frac{6}{11}$
21	$5\frac{1}{4}$	$37\frac{15}{16}$		$67\frac{1}{16}$	$66\frac{4}{7}$
20	5	$41\frac{9}{16}$		$70\frac{11}{16}$	70 --
19	$4\frac{3}{4}$	45 --		$74\frac{2}{16}$	$73\frac{11}{19}$
18	$4\frac{1}{2}$	$48\frac{12}{16}$		$77\frac{14}{16}$	$77\frac{2}{3}$
17	$4\frac{1}{4}$	$53\frac{11}{16}$		$82\frac{12}{16}$	$82\frac{4}{17}$
16	4	$58\frac{2}{16}$		$87\frac{14}{16}$	$87\frac{3}{8}$
15	$3\frac{3}{4}$	$63\frac{15}{16}$		$93\frac{1}{16}$	$93\frac{1}{5}$
14	$3\frac{1}{2}$	$71\frac{5}{16}$		$100\frac{7}{16}$	$99\frac{6}{7}$
13	$3\frac{1}{4}$	$78\frac{11}{16}$		$107\frac{13}{16}$	$107\frac{7}{13}$
12	3	$88\frac{7}{16}$		$117\frac{9}{16}$	$116\frac{4}{8}$

AA. The number of equal ſpaces in the ſhorter leg, that contained the ſame parcel of Air diverſly extended.
B. The height of the Mercurial Cylinder in the longer leg, that compreſs'd the Air into thoſe dimenſions.
C. The height of a Mercurial Cylinder that counterbalanc'd the preſſure of the Atmoſphere.
D. The Aggregate of the two laſt Columns *B* and *C*, exhibiting the preſſure ſuſtained by the included Air.
E. What that preſſure ſhould be according to the *Hypotheſis*, that ſuppoſes the preſſures and expanſions to be in reciprocal proportion.

For the better underſtanding of this Experiment it may not be amiſs to take notice of the following particulars:

1. That the Tube being ſo tall that we could not conveniently make uſe of it in a Chamber, we were fain to uſe it on a pair of Stairs, which yet were very lightſom, the Tube being for preſervations

Boyle's table indicating the reciprocal relation between pressure and volume of a gas, as given in *Spring and Weight of the Air*, 2nd edition, 1662.

and E in Boyle's published table of results. A similar set of results was obtained in expansion experiments carried out by raising a narrow tube containing a plug of air sealed in by mercury, from a wider tube full of mercury.

By 1665 a second pump design was in use, and experiments carried out to examine among other things *Rarefaction of the Air, Relation between Flame and Air* and *Hidden Qualities of the Air*. (All of Boyle's books and papers are listed in Fulton's bibliography, and may be located in the new edition of Boyle's works edited by Hunter and Davis. Both these books are listed in Further Reading.)

Boyle did work of the first order on light and colours. His treatise on *Experimental History of Colours* ranks in importance with *Spring and Weight of the Air*. It deals with a variety of phenomena including the use of a prism to resolve white light; the diagram of the experiment shows the externally reflected as well as the internally refracted rays. It included a description of experiments which showed that black bodies absorb light while white bodies reflect it, and studies of thin films. Important to chemists are the experiments in which he observed the colour changes in solutions of vegetable extracts upon change from acidic to alkaline conditions, and this established the use of acid–base indicators. The book paved the way for Newton, and many of Boyle's generalizations were adopted by Newton in his treatise *Optics*.

Boyle's treatise on *Experimental History of Cold*, published in 1665, is a milestone in science since it applies a quantitative tool, namely a thermometer, to the study of the interaction of elemental substances and mixtures. Boyle was aware of the advantages of sealed compared with unsealed thermometers, and of the necessity of a standard for cold, a fixed point of reference for which Boyle suggested the use of oil of aniseed. He advocated the use of spirit of wine (ethyl alcohol) coloured with cochineal as the liquid for thermometers because it was free from the inconvenience of freezing during his experiments. Boyle was aware of the possibility of preserving food by cold storage, of the force of expansion of water on freezing, and of the effects of temperature in physiology.

Some of his experiments display an element of humour; for example, the demonstration of making a colourless liquid boil by addition of ice, the liquid in question being concentrated sulphuric acid.

In *Mechanical Qualities*, a collection of 12 independently titled tracts which was published in 1676, he discussed the *Mechanical Origin of Heat and Cold*, and also included tracts on *Magnetism* and *Electricity*, the latter being the first tract in English on electricity.

The principle of weighing bodies in air and in water dates from the time of Archimedes, but it was Boyle who first directed attention to the importance of specific gravity measurements. References are scattered through his many works to the use of specific gravity in the analysis and characterization of materials, for example in *Origine and Virtue of Gems*, published in 1672.

His major text *Medicina Hydrostatica*, published in 1690, was the first tract in the English language on the determination of specific gravity. Despite the title the book is essentially a monograph devoted to a specialized branch of physical testing of materials. Boyle knew how to deal with materials that were less dense than water or dissolved in water. The results were expressed in decimal form, for example, for rock crystal, $2\frac{21}{100}$. Results for many materials, where purity was not a problem, showed good agreement with modern values; for example, for quartz (agate), Boyle's value of $2\frac{64}{100}$ compares well with the modern value of 2.65.

Boyle's publications on moralistic, theological and utopian writings are less well known these days, yet were significant at the time, and comprise about one third of his output. His first book, predating *Spring and Weight of the Air* by one year, was *Some Motives and Incentives to the Love of God,* published in 1659. It is known as *Seraphic Love* from the running page heading used in all the seventeenth-century editions. It was a best seller in its day and appeared in 12 English editions, and one Latin, one German and one French edition.

In the 'advertisement to the reader', Boyle gave a revealing apologia. He wrote: '... the former papers (written in compliment to a fair lady)...'. This lady has been identified by some as Ann Howard to whom the Earl of Cork had intended a match; so keen was he that a second ring was given in 1642 in addition to that given in 1641. In Robert Boyle's will more than 40 years later he bequeathed a ring to his sister, 'a ring worn for many years for a particular reason, not unknown to my dear sister', probably that returned to him in his youth. Boyle never married despite numerous attempts to provide a wife for such an eligible bachelor.

Throughout his life Boyle's interests were divided between science and theology, but he was careful not to mix one with the other, although many of the theological works have illusions and arguments drawn from natural history and science.

The Style of the Scriptures, written in 1661, was the first of his devotional works, and is a remarkable forerunner of modern higher criticism, which is usually said to have begun nearly a century later with Jean Astruc in 1753. Boyle compared the Gospels in a truly scientific fashion, commenting on various incongruities but emphasizing their essential harmony. A few months later the King appointed him as a Governor of the Company for the propagation of the Gospel among the Heathen Nations of New England and other parts of America. Boyle supported financially the translation of the Bible into numerous languages including Native American dialects, Malayan, Arabic and Irish. The production of a fresh font of type was necessary for the printing of the Irish New Testament; Joseph Moxon cast this.

Boyle achieved a great reputation among the American colonists, and they kept him informed of their affairs, and sent him news of scientific wonders and strange events. He also maintained a deal of correspondence at home and in Europe. Much of this is now available in Birch's early collected works, and in the new and comprehensive edition of Boyle's correspondence edited by Hunter, Clericuzio and Principe.

The correspondence is a key element of his writings, which contain three distinct elements or components. The first or traditional is a result of his reading a great variety of authors. The second or experimental is based on the outcomes of his own researches or those of his paid assistants in the laboratory. The third component is a mass of information and misinformation acquired as a result of conversation or correspondence with many people, navigators, travellers, 'credible persons' and the like. The extent of his reading can be judged from the references and acknowledgements in his texts; if references are omitted, reasons are given. A more detailed evaluation is not possible because Boyle's library was sold piecemeal and uncatalogued.

Reading Boyle's scientific or other works requires a considerable amount of dedication because, as expressed by Thorndyke (in the book listed in Further Reading): 'Boyle is notorious as having one of the most tiresome literary styles on record. It is diffuse, rambling, apologetic and self-deprecating and without terminal facilities.' Another detriment to the easy perusal of Boyle's writings is their over-elaborate and confusing subdivision into books, parts, essays, discourses, sections, titles, chapters, observations, notes and whatnot, to say nothing of the profuse preliminary prefaces by author and by publisher. A number of his books are written in the form of a familiar discourse, and informal and unstructured conversations, which adds nothing to the clarity, or rapidity of transmission of thought.

However, among all the prolixity and verbiage occur some strikingly prescient and illuminating opinions such as those on the nature of an element, on heat and on sound. On an element: 'And, to prevent mistakes, I must now advertize to you, that I mean by Elements, as those Chymists that speak plainest do by their Principles, certain Primative and simple, or perfectly unmingled bodies; which not being made of other bodies, or of one another, are the ingredients of which all perfectly mixt bodies are immediately compounded and into which are ultimately resolved.' Of the mechanical origin or production of heat: 'The nature of it [heat] seems to consist mainly, if not only, in that affectation of matter we call local motion.' And on sound: 'Sound consists of an undulating motion of the air.'

Robert Boyle died on 31 December 1691, seven days after his beloved sister Katherine. He was buried close to his sister in the chancel of the church of St Martin-in-the-Fields, Westminster on 7 January 1692. Surprisingly no memorial appears to have been erected to Robert Boyle or to his sister in the church where they were buried. When the old church was demolished around 1720 no systematic record was made of the disposal of the remains of bodies interred there. The present church was built between 1720 and 1724; it contains no memorial to Robert Boyle, or to his sister; nor does it contain the remains of Robert Boyle, whose final resting-place is unknown.

Boyle is commemorated by two lecture series. The first series, founded by Boyle himself under the terms of his last will and testament, are an annual series of lecture sermons that commenced in 1692; Richard Bentley gave the first lecture sermon. The Oxford University Junior Scientific Club instituted the other series in 1892; Sir H W Acland gave the inaugural lecture. Three medals carry Boyle's name; the oldest medal, that of the Royal Dublin Society, was founded in 1899; the first recipient was G J Stoney*. The next, the Robert Boyle Gold Medal of the Analytical Division of the Royal Society of Chemistry, restricted to overseas candidates, was first awarded in 1982 to Sir Alan Walsh. The most recent, the Boyle-Higgins Gold Medal of the Institute of Chemistry of Ireland was first awarded in 1990 to D T Burns.

Further Reading

The Life of the Honourable Robert Boyle FRS, R E W Maddision (Taylor and Francis, London, 1969).

The Works of the Honourable Robert Boyle, T Birch (ed) (A Millar, London, 1744, 5 volumes); a second edition of 6 volumes, edited by J and F Rivington *et al*, was published in 1772.

A Bibliography of the Honourable Robert Boyle, Fellow of the Royal Society, J F Fulton (Clarendon, Oxford, 2nd edition, 1961).

The Works of Robert Boyle, M Hunter and E B Davis (eds) (Pickering and Chatto, London, volumes 1–7, 1999 and volumes 8–14, 2000).

Robert Boyle and Seventeenth-Century Chemistry, M Boas (Cambridge University Press, 1958); reprint (Kraus Reprint Co, Millwood, New York, 1976).

Robert Boyle (1627–1691): A Foundation Stone of Analytical Chemistry in the British Isles, D Thorburn Burns, published in *Analytical Proceedings*. Part I. *Life and Thought* **19** (1982) 222; Part II. *Literary Style, Specific Contributions to the Principles and Practice of Analytical Chemical Science* **19** (1982) 288; Part III. *American and Dutch Connections* **22** (1985) 253; Part IV. *Robert Boyle's Determination of Iron in Tunbridge Water: The Earliest Quantitative Colorimetric Analysis?* **23** (1986) 75; Part V. *Hungarian Mines, Minerals and Mineral Waters* **23** (1986) 349; Part VI. *Contributions to the Early Study of Luminescence* **28** (1991) 362.

Opticks: or a Treatise of the Reflections, Refractions, Inflexions and Colours of Light, I Newton (Smith and Walford, London, 1704).

The Correspondence of Robert Boyle, M Hunter, A Clericuzio and L M Principe (eds) (Pickering and Chatto, London, 2001, 6 volumes).

A History of Magic and Experimental Science, L Thorndyke, Volume VIII: *Seventeenth Century* (Columbia University Press, New York, 1958).

William Molyneux

1656–1698

Patrick Wayman and Norman D McMillan

William Molyneux was born in Fishamble Street in Dublin on 17 April 1656. During the 25 years between 1661 and 1685, following the Restoration of the English crown to Charles II, Dublin experienced considerable prosperity, in contrast to the strife of the preceding periods of war and the Commonwealth. The population of the city increased from 20 000 to 60 000, and Dublin became Britain's second city. The long-established families of the Pale flourished, and Samuel Molyneux (1616–1693), William's father, and his capable wife Margaret (née Dowdall, d. 1700), were able to purchase a country estate, Castle Dillon in County Armagh, in 1664, and a substantial town house in 1665. It was in this atmosphere that Samuel's eldest son, William, was educated sufficiently to enter Trinity College as a fellow commoner, a relatively privileged status, in 1671, when he was 15.

Recognizing the cooperative nature of scientific discovery, he was insistent on the importance of establishing a society in Ireland equivalent to the Royal Society of London, and in 1684 was principal founder of the Dublin Philosophical Society or DPS, which brought him in contact with the principal figures of the day connected with Trinity College Dublin. The society only lasted until 1708, ten years after his death, but its example was followed in the foundation of the Dublin Society in 1731 (Royal Dublin Society from 1820), and of the Royal Irish Academy in 1785.

In addition, William's son Samuel, who was a government official with responsibilities for artillery and gunnery, and who was concerned with the rapid development of engineering and physical science, was highly involved in the DPS after William's death. Other family members of the Molyneux family, such as William's brother, Sir Thomas Molyneux (1661–1733), first Baronet, and his descendants continued to support such institutions. (The baronetcy died out with the 10th baronet in the 1890s.) In fact Sir Thomas was one of the most important figures in the foundation of the Dublin Society; here the work of the DPS was carried forward, and continuity was provided with modern institutions.

Molyneux was dissatisfied with what he considered the scholastic curriculum in Trinity and, according to Simms (his biographer—see

William Molyneux.

Further Reading), was very greatly attracted to 'Lord Bacon's method and those prescribed by the Royal Society'. Molyneux was obviously a remarkable student; leaving Dublin immediately after graduation in 1674, he travelled to London, where he was elected a Fellow of the Royal Society. While the Restoration created some difficulties for the members of the Royal Society, the conditions that prevailed in Dublin for the enthusiasts for the so-called 'new learning' were much more extreme. After Molyneux returned to Dublin, the situation rapidly developed into one of open war.

The 'Dublin Philosophical Society for the Improvement of Natural Knowledge, Mathematics and Mechanicks' was founded in October 1683, the founder being Molyneux, with the assistance of the Provost of Trinity College Dublin, Narcissus Marsh (1638–1713). The society held meetings in Trinity, and had a room in College and a laboratory in Crow Street with a telescope that was purchased for £20. In 1686 the Herbal Gardens were established for the Society by a member called Huntingdon. The fledgling society grew to some 33 members in its first year, but soon ran into problems of political repression from the Catholic Lord Lieutenant Richard Talbot, Duke of Tyrconnell (1630–1691).

In 1686 Talbot personally blocked the plan to obtain a Royal Charter for the society, which the members had hoped would secure for them physical protection, following the example of the London sister society. Their anxiety over their physical safety proved to be well founded, as the Royalists occupied Trinity following James' landing at Kinsale in 1689, though Molyneux and seven other members of the DPS had already wisely decamped from Dublin in 1687. The Molyneux brothers both fled to Chester where William began the work on his book *Dioptrica Nova.* In Trinity, the Royalists expelled both the staff and students of the university, and for good measure destroyed all Molyneux's astronomical equipment.

After the defeat of James, Molyneux returned to Dublin to become the central figure in the new government. Consequently, when the DPS was re-established in 1693 it was without Molyneux's active involvement. The membership grew rapidly from 16 initially to become 49 by the end of that year. For various reasons, though, formal meetings of the society ended in 1697, and, probably because of Molyneux's personal problems over his book, *The Case of Ireland*, which will be discussed below, the following year the society came to an end. Molyneux died on 11 October 1698 in Dublin from an aggravation of the kidney complaint which had caused him problems all his life. It can be surmised that the principal causes of this second failure of the DPS were the dispersal of members, and their elevation in government and church. The DPS was subsequently re-established in 1706 by his son Samuel; this third period was perhaps most notable for the membership of the philosopher George Berkeley (1685–1753).

It is not possible to over-state the importance of Molyneux's work in establishing the DPS, but equally important were his own many diverse and seminal personal contributions to science and astronomy, and optics in particular. William's education at Trinity led him towards mathematics, and, in his twenties, he prepared translations of Descartes' *Meditations*, which remained unpublished. Maintaining for a considerable time an amiable correspondence with Flamsteed at Greenwich, he became knowledgeable on astronomy, and the related sciences, especially optics.

He was involved in controversies with Hooke and Hevelius, both leading scientific philosophers in Europe. In 1685 he undertook a commission from the editor of the *Philosophical Transactions* to write a Latin translation of a review of Hevelius' book *Annus Climactericus*, which had been published in Danzig in that year. The book had been attacked by Hooke in the Royal Society for its statements concerning the superiority of plain sights in telescopes. Molyneux was attempting to smooth ruffled feathers, but actually succeeded in causing Hooke annoyance, for calling his book *Animadversions on the First Part of [Hevelius's] Machina Coelestis*, which Hooke had published in London in 1674, a 'pamphlet'.

Molyneux worked on a number of astronomical topics. His astronomical work, like that of so many of his generation, had a religious inspiration. He was concerned to try to convince himself of the existence of a Supreme Being, from his studies of the rational nature of the solar system.

In 1683 he studied the question of why the apparent magnitude of the sun near the horizon was greater than at its zenith, and he was extremely interested in all issues relating to telescope sights. Also in 1683, he presented to the DPS a paper on double vision, and in 1684 he cooperated with various scientists in London on observations on solar and lunar eclipses. He became involved in debates over the action of lenses in 1685, and this work eventually led to his major research field of astronomical optics. He designed a new telescope dial of his own design, and in 1686 published *Sciotherium Telescopicus* to popularize his invention, though the dial unfortunately never came into widespread use.

He continued with capable work in instrumental science and, in 1692, wrote the first account of optical science in English—*Dioptica Nova*. The book was to become a standard text, partly because it was written, after careful consideration of the issue, in English, rather than Latin. The book was lavishly and discerningly illustrated; Edmond Halley helped Molyneux revise the text of this book and it thus attained a very high standard. Furthermore, the Englishman allowed Molyneux to reproduce for the first time his famous theorem on the finding the foci of optical glasses, making it then of some additional research interest. Thus the writing of this work brought these two substantial historical figures of science together. The book was published in two parts with the second part dedicated to Henry Osborne, a friend of Robert Hooke and a surveyor from Meath.

Molyneux was the weather diarist of the DPS until 1684 when St George Ashe replaced him. This work required the regular recording of temperature, atmospheric pressure, rainfall, points of the wind, and humidity, using instruments that were still in the course of steady development. This obviously gave considerable scope for instrumental improvements, which indeed featured in papers written by Molyneux, and also other members of the DPS. In 1685, and in collaboration with Marsh, he developed what was known as the Dublin hygroscope, developing a design of Hooke.

Molyneux took part in a considerable range of more miscellaneous scientific tasks. He was involved in the work on models for the optimum design of the double-bottom boat of William Petty* (1625–1687); these models were tested, as in modern times, in large water tanks. In 1685 he toured the continent and studied fortification and the state of inland navigation, and he also produced, in collaboration with Flamsteed, a paper on high water at various places in Dublin which was published in the *Philosophical Transactions* in 1686. Unfortunately, though, Flamsteed was to fall out badly with Molyneux through his unwarranted criticism of *Dioptica Nova*.

We now discuss his scientific, political and philosophical legacy. Molyneux was a representative of what has been referred to as 'second-stage Protestants'; these were men who opposed the enthusiasms of the Puritans. Conveniently they found that the Baconian philosophy, which was so central to the supporters of the Commonwealth, was malleable enough to be accommodated in the new situation after the Restoration of the monarch. While Molyneux supported the gains of Parliament in the Act of Settlement, he was generally tolerant in his views, and, indeed, in 1685, he formally welcomed the new Catholic king, but the events of the monarch's short reign soon decided Molyneux, and the DPS in general, to come down firmly in favour of the Dutch William.

In truth, Molyneux was a very typical Baconian. In letters to Halley he derided the party inside the Royal Society which rejected all kinds of 'useful knowledge', and in his book *Sciothericum Telescoicum*, he made large claims for science, whose main task he saw as the 'advantage of mankind'. The DPS was sympathetic to Baconian cooperative efforts, as demonstrated by Molyneux's leading role in arranging contributions for Moses Pitt's projected *English Atlas*. This wearisome task of organizing the collection of facts related to Ireland for this project was supervised by Molyneux with great success over the period 1679–1682, and indeed appears to have become the main activity of the DPS meetings during that period. However, the project collapsed with the arrest of Pitt for debt, and sadly much of the 'rude (unordered) material' gathered by the DPS was eventually burned.

Molyneux had a greater grasp than any other DPS members of the nature of the new approaches to physics which were being pioneered by Galileo, Torricelli and Newton. He himself translated from Italian the third and fourth days of Galileo's *Dialogues concerning the Two Chief World Systems* and Torricelli's *De Motu Gravium Naturalitur Descendentium et Projectorum*, and thus put at the disposal of the DPS two of the most important works of the day. Molyneux was amongst the first to comprehend the enormous importance of Newton's work after he received advanced copies of sections of *Principia* from his friend Halley in 1687. In 1697 he attempted to find a mathematician to write a popular English account of Newton's work, but he died before anything could come of his efforts.

After the pivotal year for the history of Ireland of 1690, William Molyneux interested himself in political and economic problems, and in his last year, at the age of 42, having taken his cue partly from the historical works of his father-in-law Sir William Domvile, he published *The Case of Ireland being Bound by Acts of Parliament in England, Stated*, urging equal rights for Ireland under the Crown. It was his intention to uphold traditional rights for the Irish parliament, and to resist restrictions on Irish trade. He was influenced by his friend and collaborator John Locke in recognizing representative government as a natural right, independent of historical precedent. Due perhaps to special personal interests, his book did not secure the recognition that many, such as (Dean) Jonathon Swift, thought it deserved. A second edition was published in Dublin in 1749 and others were published in London in 1770, and Dublin in 1773, bringing in the interest of the American colonists. These led to partial reforms, which were largely obliterated by the Act of Union in 1800.

Although this work is little more than a pamphlet, its long-term influence is difficult to overestimate. Molyneux wrote to Locke that he had treated this nice subject with caution as 'I cannot justly give offence'. In this, he was very mistaken. The English Parliament ordered the book to be burned for they saw immediately its dangers. It is here that the ideas are set forward for the 'patriot' movements that later rose, despite the fact that the work was dedicated to William III. The economic slogan 'No taxation without representation' first finds expression here.

William's wife Lucy (née Domvile), who died in 1691, suffered from blindness; this defied all attempts at cure, and made the family life difficult. William's resultant interest in the problem of blindness led him to propose to John Locke in 1688 the celebrated 'Molyneux problem', largely unresolved to this day. The problem inquires whether a person who was blind from birth, and who had become accustomed by touch to shapes such as a sphere or a cube, could, on gaining the gift of sight, immediately distinguish between them visually. It was Locke who popularized the problem. The friendship between the two men was a close one, and Locke was left a legacy in Molyneux's will. After his death their correspondence was published as *Some Familiar Letters between Mr Locke and Several of his Friends* (London, 1708) by his son Samuel.

When Molyneux died, he left a profound and lasting heritage. He was perhaps the single most important figure in the history of Irish science, and one of very great political significance. The Royal Dublin Society, the Royal Irish Academy, the Institution of Engineers of Ireland, together with numerous other Irish professional societies such as those in mathematics, statistics, political economy, geology, botany, chemistry, physics and other disciplines trace their origins directly to the Dublin Philosophical Society, and have at various times acknowledged the Society or Molyneux as their inspiration.

Further Reading

William Molyneux: Astronomer and Natural Philosopher, Alan Gabbey in: *Some People and Places in Irish Science and Technology*, Charles Mollan,

William Davis and Brendan Finucane (eds) (Royal Irish Academy, Dublin, 1985), pp 10–11.

William Molyneux of Dublin, J G Simms (Irish Academic Press, Blackrock, 1982).

The Common Scientist in the Seventeenth Century: A Study of the Dublin Philosophical Society 1683–1708, K Theodore Hoppen, (Routledge and Kegan Paul, London, 1970)

Prometheus's Fire: A History of Scientific and Technological Education in Ireland, N D McMillan (Tyndall, Carlow, 2000). For a detailed discussion of the specific fruits of the DPS see pages 92–105, but generally the book details throughout the actual histories of various facets of education and professional activity that are a direct outgrowth of the DPS.

A New History of Ireland, volume 3, T W Moody and F X Martin (eds) (Clarendon, Oxford, 1976).

Nicholas Callan

1799–1864

Niall E McKeith

Nicholas Joseph Callan was a priest and scientist; from 1826 until his death in 1864 he was Professor of Natural Philosophy at Maynooth College, which had been founded in 1795 by the Irish Parliament for the education of Catholic clergy. He was a pioneer in electrical science, constructing great batteries, electromagnets and electric motors, his greatest achievement being the invention of the induction coil in 1836. His great coil of 1837 was capable of producing fifteen-inch sparks, thus producing hundreds of thousands of volts from a source of five or six batteries.

Credit for his discoveries and inventions has often been given to others; in particular that for the induction coil has often gone to the German inventor Heinrich Ruhmkorff. However, expert opinion from the Earl of Rosse* shortly after Callan's death, to the famous expert on electricity and electronics, J A Fleming, in the 11th edition of the *Encyclopaedia Britannica,* has given Callan full credit for its invention. In 1999 the Irish Branch of the Institute of Physics erected a plaque on the site of his laboratory, and An Post, the Irish Postal Service, brought out a stamp as part of their millennium collection, in recognition of his contributions in the field of electricity

Callan was born on 22 December 1799 at Darver, between Drogheda and Dundalk in County Louth. The Callans were a well-to-do family of considerable repute in the district. They farmed extensively and, in addition, were bakers, maltsters and brewers, and it was Callan's private wealth which enabled him to fund his research. During the famine period of the 1840s, it would also allow him to devote the whole of his professorial salary to relief projects. He was the fifth son of a family of seven. He received his early education at Dundalk Academy; he was then sent to Navan Seminary to do his initial preparation for the priesthood, and in 1816 he entered the National Seminary at Maynooth. Here he excelled in all subjects, including theology, ecclesiastical history and canon law, but he was particularly aroused by the lectures of Dr Cornelius Denvir on Experimental and Natural Philosophy. Denvir was interested in magnetism, and his influence largely determined the direction of Callan's future researches.

Nicholas Callan (courtesy Maynooth College).

Callan was ordained priest on 24 May 1823; he was just 23. It seems that he had been left a substantial legacy on condition that he did *not* become a priest before the age of 26; this he was willing to forgo. On completion of his studies in 1824 he was sent to Rome, where he attended lectures in the Sapienza University and obtained a Doctorate in Divinity in 1826. During his stay in Rome he became acquainted with the works of Galvani and Volta and his interest in magnetism discovered under Denvir was reinforced. On the resignation of Dr Denvir, his former Professor, he then applied for the Chair of Natural Philosophy back in Maynooth, and was appointed in 1826.

There were ten Professorial Chairs in the College. The Chair of Natural and Experimental Philosophy, to which Callan was appointed, embraced mathematics, mechanics, astronomy, and such experimental sciences as existed at that period, and as such is the oldest department of Experimental Physics in Ireland, apart from that at Trinity College Dublin.

It was in the field of electricity that Callan showed his inventive genius. Apart from a period in the 1840s, it does not seem that he felt any conflict between his roles as priest and man of science. Rather he felt his intellectual curiosity was worthily directed towards investigation of God's creation. In addition he had a vision of electricity in the service of man, hoping to see it as a cheap and convenient source of light and power. In almost every way his vision has become reality, and Callan made several of the important steps towards achieving this.

To describe his researches, it will be of advantage not to consider them in chronological order, since many of his areas of research were advanced simultaneously. Instead they will be considered under various headings, all closely interrelated: electromagnets and the induction coil; electric motors; and batteries.

With the advent of batteries, it was soon discovered that a coil of wire carrying a current became magnetic, and also that, when the coil was wrapped around an iron bar, it rendered the bar magnetic. In England, Sturgeon and Faraday constructed powerful electromagnets, and so did Callan, some of whose are on display in the National Science Museum at Maynooth. The great horseshoe magnet stands nearly 6 feet (2 m) in height, and the diameter of the poles is about $2\frac{1}{2}$ inches (6.35 cm). It weighs approximately 210 pounds (95 kg). James Briody, the local village blacksmith, did the ironwork, while Callan wound heavy copper wire on the poles. This was in the late 1820s, long before insulated wire came upon the market, so Callan had to insulate it by wrapping tape around it. This huge electromagnet of Callan's had a lifting power of some two tons when supplied with current from his battery.

In his invention of the induction coil Callan was influenced by the work of Henry and Page in America, who independently discovered self-induction by interrupting the current flowing through a spiral of copper ribbon. They felt a slight shock if the ends of the ribbon were held in their hands. Callan introduced copper wire in place of the ribbon and, more importantly, he wound the wire, insulated with tape, on an iron core. In this way he obtained shocks of greater intensity. In one of his first coils he wound two lengths of copper wire, each 200 feet (61 m) long, around a straight iron bar of one-inch (2.54 cm) diameter. The two windings were joined in series, but the current was sent only through one winding, whilst the shock was taken from the full 400 feet (122 m) of coil; i.e. it was an *autotransformer* arrangement. Used with one of his cells, this apparatus gave shocks of great intensity.

In fact, he used this as one method of testing the power of his batteries. He persuaded his students to take shocks from it, and from the reactions of the students he judged the power of the cell; he recorded that with 14 Wollaston cells, the shock was so strong that the person who received it felt the effects for some days! He also added that, with 16 batteries, nobody could be persuaded to take the shock. There is a story that one student named William Walsh, who later became Archbishop of Dublin, was rendered unconscious by a shock. Eventually the College authorities insisted that he stop using students in his experiments.

Callan was the first to recognize *(circa* 1836) that the intensity of the shock depended on the rapidity of the break in the circuit. He constructed a device, made from the escapement mechanism from a grandfather clock, for interrupting the current extremely rapidly (2000 to 3000 times per minute). He attached a crank handle to the escapement cogwheel and a thick copper bar to the rocker, which dipped in and out of cups filled with mercury. On rotating the crank, he got more than 50 interruptions a second. This device, which he referred to as a 'repeater', is still preserved in the College Museum. It focused attention on the significance of the rapid change in magnetic flux.

In his next experiment, Callan used a primary coil of 50 feet (15 m) of thick copper wire and a secondary coil of 1300 feet (396 m) of fine wire. In preliminary trials he sent the battery current through the primary, and took the shock from both the primary and the secondary coils in series. His next step was to separate completely the primary and secondary coils. Thus Callan discovered for himself the principle of the step-up transformer of modern high-voltage electricity. The secondary current was sufficient to strike an arc between the carbons in an arc lamp or, more dramatically if less pleasingly to today's tastes, to electrocute a large fowl.

As a result of these experiments, by 1836 he had all the elements of the modern induction coil, apart from the condenser (or capacitor, in today's terms). The use of the latter was suggested some 17 years later by Fizeau, its function being to reduce the sparking in the primary circuit breaker. Callan sent a replica of this apparatus to Sturgeon in London, and Sturgeon showed it to members of the London Electrical Society in August 1837. Many induction coils were then constructed by Sturgeon himself and several others, all clearly giving credit to Callan.

Callan now built the 'medium' coil, and then, in 1837, the 'giant' coil; both are preserved at Maynooth. The overall length of the latter was about 5 feet (1.5 m). The primary coil consisted of thick copper wire, tape-insulated, wound around a core of iron rods about 40 inches (1 m) in length. There were three secondary coils, said to contain about 150 000 feet (45.7 km) of fine iron wire, all hand-insulated.

In his address to the British Association meeting in Dublin in 1857, Callan made the following statement:

'It is now more than twenty years since I discovered the method of making the induction coil, or a coil by which an electric current of enormous intensity may be produced with the aid of a single galvanic cell; a coil which is now to be used for the working of the Atlantic Telegraph. Mr Faraday was the first to develop the laws of electrical induction, but he did not discover the method of making a coil by which a current of very great intensity may be obtained by means of a very small battery. This was first discovered in Maynooth College in 1836. In the summer of 1837, I sent to the late Mr Sturgeon a small coil, which he exhibited at a meeting of the Electrical Society in London and from which he gave shocks to several of the members.

'This was the first induction coil of great power seen outside the College of Maynooth. The first notice of the discovery of the coil is to be found in a paper of mine published in the *London Philosophical Magazine* for December 1836. In April 1837, I published in Sturgeon's *Annals of Electricity* a description of an instrument which I devised for producing a rapid succession of electrical currents in the coil by rapidly making and breaking communication with the battery [i.e. his repeater]; thus before April 1837 I had completed the coil as a machine for producing a regular supply of electricity. From 1837 to the end of 1854, my attention was directed to other matters.'

To his desired end of electricity in the service of man, Callan constructed motors driven by electricity obtained from his battery. They were essentially simple in construction and operation, the rotors consisting of a series of iron bars or plates mounted on the periphery of a wheel, and the end of the axle carrying a small wheel with pointed teeth which made contact with a stationary spring strip. When this contact was made, the current flowed to energize the stationary electromagnets which attracted the bars. Just as the bars passed over the poles of the magnets the teeth lost contact with the strip and the current was interrupted. The momentum kept the machine rotating until the next tooth made contact, when the process was repeated. He experimented with a number of motors of this type, and had hopes of using them to electrify the railway line from Dublin to Dun Laoghaire (formerly Kingstown). An electric motor of the type described by Callan was manufactured in Dublin in the late nineteenth century by Yeates and Co.

Callan estimated that an electric motor as powerful as the steam engines then in use could be built for £250. It would weigh less than two tons, and could be maintained at an annual cost of £300, which was a quarter of the cost of steam power. He designed an engine intended to propel a carriage and load at a speed of 8 miles an hour, but he met with a number of snags. The batteries were spillable and unwieldy, electromagnetic action is powerful only at a short distance, and large magnets interfere with one another. In time he learnt that laboratory-scale tests do not always apply to large-scale work, and in the end he was forced to abandon the idea.

A steady objective in Callan's research was the cheap production of electricity, at a time, of course, when the dynamo had not been invented. Frictional machines produced electricity of high voltage, but they could not drive an electromagnet as their current was low. Voltaic cells on the

Yeate's motor based on Callan's design (courtesy Maynooth College).

other hand could produce electricity with high current, but not at high voltage, although this could be increased by connecting cells in series. Callan experimented with the Wollaston cell, a zinc/copper/sulphuric-acid arrangement, and also with the 'improved' Wollaston cell, also known as the 'double copper' in which the copper plate had the shape of a U with the zinc plate in the centre; this effectively doubled the useful area of the zinc plate. Callan improved on this by using a copper container which acted as the positive plate, and thereby eliminated the cost of a porcelain or glass container, and made the whole thing cheaper to construct.

It had been established for some years that the quantity of electricity from a voltaic cell increased with the area of the plates, and the intensity with the number of cells in the battery. Callan set out to obtain large quantities of electricity, and his first step was to use plates of much greater area than usual. In a paper in the *Philosophical Magazine* in 1836 he described 'a very large battery, lately constructed for the college at Maynooth'. This consisted of 20 enormous zinc plates, each 2 feet square, and 20 copper cells, each large enough to contain a zinc plate, with clearances of about a quarter of an inch between zinc and copper surfaces. The plates were covered with woven nets of hemp to help prevent contact with the copper, and they were raised and lowered by means of a windlass. It required nearly 30 gallons of acid to fill the whole battery. All the zinc and all the copper plates were joined together so as to give a total effective surface area of 160 square feet.

According to Callan: 'So enormous is the quantity of electricity circulated by this battery when all the zinc and copper plates act as a single circle, that on one occasion, after having acted without interruption for more than an hour, it rendered powerfully magnetic an electromagnet on which were coiled 39 thick copper wires. On the fifth day it was tried: after having been in action without interruption for more than two hours this battery melted very rapidly a platina wire 1/30th of an inch thick, and deflagrated in a most brilliant manner copper and iron wire about 1/12th of an inch thick'.

His main achievement was the cast iron battery. In 1847 he published in the *Philosophical Magazine* a paper entitled 'On a new voltaic battery, cheap in its construction and use and more powerful than any battery yet made, and also on a cheap substitute for nitric acid of the Grove's platinum battery'. He first substituted lead for the platinum. Although the platinum cell was initially twice as powerful as the lead cell, at the end of $3\frac{1}{2}$ hours the lead was twice as powerful as the platinum. The use of cast iron was also successful. Nitric acid, the fluid used in these cells, was very expensive, and also attacked the gilding and platinizing which he sometimes used on the lead plates. He found that by adding nitre (potassium nitrate) to sulphuric acid, and diluting it with an equal volume of water, he got a much cheaper mixture, which was even better than that using the concentrated acid.

He next decided to convert the existing Wollaston batteries into cast-iron ones. He built a total of 577 cells, with a combined area of zinc of some 96 square feet ($9\,m^2$). Sixteen gallons of sulphuric acid and fourteen gallons of nitric acid were required. This was reckoned to be the world's largest battery at the time, being at least twice as large as that constructed at the École Polytechnique on Napoleon's orders. E M Clarke subsequently manufactured a 'Maynooth Battery' of this sort commercially at the Adelaide Gallery of Practical Science, 428 The Strand, London. On 7 March 1848, before a large audience in the college, Callan demonstrated the power of his battery. A 5-inch (13-cm) arc of blinding light was caused to appear between brass and copper terminals, and a large turkey placed in the circuit was instantly electrocuted!

Through this work Callan was led into metallurgy. In his choice of cast iron for his battery, he had been influenced by the fact that it withstood the action of nitric acid almost as well as platinum or gold, and was, of course, much cheaper. He discovered that iron coated with a mixture of lead and tin was much more resistant to attack than iron coated with zinc alone, as in the more usual galvanizing process. The addition of a little antimony gave a further great improvement. He

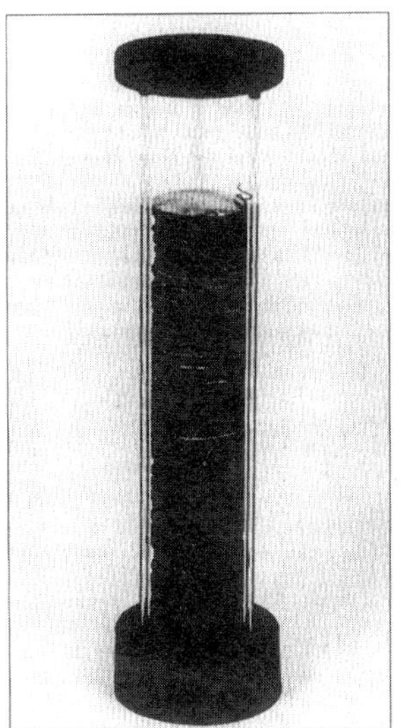

Anti-clockwise from top, Leclanché cells, voltaic pile and Maynooth battery (courtesy Maynooth College).

patented this process in 1853, and Maynooth Museum still possesses the elaborate patent document bearing the seal of Queen Victoria. The patent was entitled 'A means of protecting iron of every kind against the action of the weather and of various corroding substances so that iron thus protected will answer for roofing, cisterns, baths, gutters, pipes, window-frames, telegraph-wires for marine and various other purposes'.

Callan pursued other investigations too. He usually measured the currents supplied by his batteries by the lifting power of electromagnets, and by the heating effects produced, but he also devised a type of galvanometer which was later widely used by Stewart and Magee, and is sometimes called by their names. He experimented with various forms of electric lighting, from arc lights to limelights. By electrolysing acidulated water he produced hydrogen and oxygen, which gave him an oxyhydrogen torch with which he heated up a block of lime, producing an intensely brilliant light. He hoped that this would serve for lighthouse beacons, but practical difficulties militated against its use; hydrogen and oxygen can, of course, form a highly explosive mixture, and Callan described one occasion on which an explosion shattered the vessel in which the gases were contained.

Another interesting device was Callan's point-and-plate valve. He had discovered that the secondary voltage from his induction coil was alternating in direction. He discovered, or found by chance, that if he interposed a plate and point spark gap in the circuit, it behaved in an asymmetrical manner as far as the flow of current across the gap was concerned. This device, the first simple rectifier, was later used in the earliest X-ray apparatus, as was the induction coil.

Much of Callan's scientific apparatus is now on display in the National Science Museum at Maynooth, which is open to the public. The museum can be accessed at its website at www.may.ie/museum/

The 1830s were probably Callan's most active decade for electrical research. In the 1840s, under the influence of other Maynooth Professors, much of his effort turned to creation of devotional literature, especially translation of the works of St Alfonso Liguori. The effort taken by this self-imposed task, added to the constant demands of teaching, and endeavouring to improve the facilities available for physics in Maynooth, took their toll. Between 1849 and 1851 Callan's health failed and he spent time in Continental spas. He was not at full strength for the rest of his life, but worked at his teaching and his research.

Nicholas Callan died in 1864 and is buried in the cemetery in the grounds of St Patrick's College, Maynooth. Beside him, as was his wish, is buried Rev M T Casey, Emeritus Professor of Chemistry, who was the curator of the museum from 1975 until his death on Christmas Day 1997. The above article is principally based on articles he published, in particular that of Nicholas Callan—Priest, Professor and Scientist published in *IEE Proceedings*, vol 132 in 1985 from which most of the material was gleaned, revised and re-edited.

Further Reading

Nicholas Callan: Priest-Scientist, 1799–1864, P J McLaughlin (Clonmore and Reynolds, Dublin, 1865).

The Scientific Apparatus of Nicholas Callan and other Historic Instruments, R Charles Mollan (Samton, Maynooth and Blackrock, 1994).

William Parsons, Earl of Rosse

1800–1867

Bob Strunz

If there is one single invention that has inspired mankind to scientific endeavour, it is perhaps the telescope. From the time of Galileo, right through to the space-telescopes of the present day, astronomers have always been seeking to build bigger and more powerful instruments to enable them to see deeper into space.

The characteristic that distinguishes this branch of science is the fact that there is a virtually unlimited range of new discoveries to be made. All that is needed is observational skill and a powerful telescope. This chapter tells the story of how an Irishman, the third Earl of Rosse, an 'amateur' by his own admission, profoundly changed our understanding of the universe by building a telescope that even by today's standards was enormously powerful. At the time, it was the most powerful instrument ever made, not just by a small margin; it was four times more powerful then the next biggest instrument. It gave Lord Rosse the capability of seeing four times deeper into space than any person had ever done before. The magnitude of the achievement is underscored by the fact that it was the world's largest telescope for 70 years, from 1845 to 1915.

As a result of this achievement, Lord Rosse received many honours. He became a Fellow of the Royal Society in 1831, and was President between 1849 and 1854. He received a Royal Medal from the society in 1851. In 1862 Lord Rosse became Chancellor of Trinity College Dublin.

Telescopes have the ability to perform two distinct tasks. In the first place, they are capable of magnifying distant objects in order to make them visible to the observer. Secondly, they have the ability to gather light. In this case, they are not necessarily magnifying an object but they are effectively making it brighter and therefore visible to the human eye.

The planets in our solar system are generally very bright and thus, for their study, the key requirement for a telescope is that it is capable of magnifying their images to an extent that permits the observer to distinguish features on their surface. With a sufficiently high magnification instrument, the astronomer can distinguish features such as craters on our moon, the polar ice caps of Mars or the rings of Saturn.

William Parsons, Third Earl of Rosse (courtesy of the Royal Society).

When we turn to objects outside our solar system, however, the requirement changes from that of magnification to that of light gathering. Even the closest of these objects is so far away from us that it is not possible to magnify it to an extent that will allow us to distinguish features on its surface. The farther we move into deep space, the fainter the objects we are hoping to see become. The ability of an instrument to gather smaller and smaller amounts of light and to focus that light accurately on to the eye of the observer becomes the critical factor, rather than its ability to magnify.

It was the Italian scientist Galileo who made the telescope famous. His first instrument was constructed in 1609, and in that year alone he observed the moon and discovered four satellites of Jupiter. Galileo's telescope was composed of a system of glass lenses and suffered badly from the effects of light passing through these lenses. Galileo did not understand why this was the case, as an understanding of the behaviour of light had not yet been developed.

In 1672, Isaac Newton published an explanation of what happens when light passes from one medium to another, for example from air to

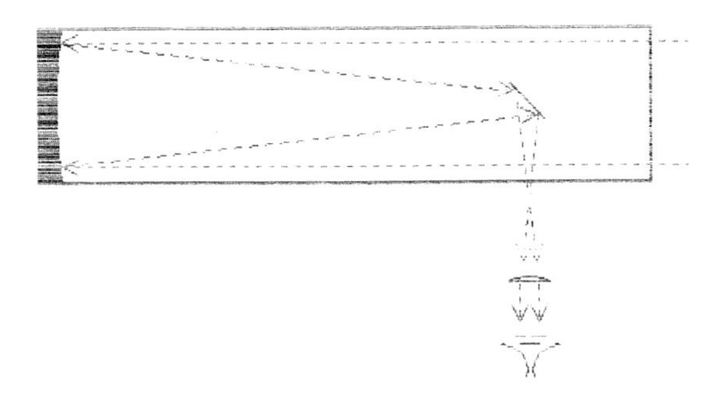

Newtonian reflecting telescope.

glass and back to air, as it goes through a lens. Newton showed that white light is a mixture of coloured light and that every colour is bent (refracted) by the transition between materials. The colours are not, however, bent by the same amount. As a result of this, a lens has a tendency to decompose light into its constituent colours and each comes to a focus at a different point. This effect is very familiar to us as it occurs when sunlight passes through raindrops to form a rainbow.

This effect, which is known in lenses as 'chromatic aberration', results in an image which is surrounded by circles of different colours; this is most undesirable in an astronomical telescope as it blurs the image.

In addition to this problem, it is also the case that light is severely 'attenuated' by passing through a glass lens. That is to say, the glass absorbs some of the light, and therefore reduces the amount that can get through to the human eye. In a situation where there are large amounts of light available, observing the full moon for example, this is not a problem, but when the observer is trying to see very faint objects, the less attenuation and the more light that reaches the eye, the better.

Newton decided to try using a mirror in place of a lens; he cast a two-inch disc of speculum metal, an alloy of copper and tin, and ground it to a spherical shape. This mirror was mounted at the bottom of a tube and caught the reflected rays on a 45° secondary mirror that reflected the image into a convex lens outside the tube. Newton's instrument was sent to the Royal Society, where it caused a sensation; it was the first working reflecting telescope. But the effort ended there. Others were unable to duplicate the curvature of his mirror and the reflecting telescope therefore remained a curiosity for decades.

In the second and third decades of the eighteenth century however, the reflecting telescope became a reality. By the middle of the century, reflecting telescopes with primary mirrors up to six inches in diameter had been made. In the second half of the eighteenth century, William Herschel successfully constructed a reflecting telescope with a parabolic mirror of 48 inches diameter.

The ability of the Newtonian telescope to gather light is related to the surface area of its mirror, and therefore it increases approximately as the square of the mirror's radius. In other words, the larger the mirror, the more light it will gather, and a six-inch diameter mirror is capable of gathering nine times more light than a two-inch one. Small increases in diameter give rise to large increases in light-gathering ability. This explains why astronomers became engaged in a race to build bigger and bigger telescopes.

The stage was now set for an individual of ability and financial means to attempt a really large mirror. It was Sir William Parsons, third Earl of Rosse, who confronted this challenge at his home in Birr Castle, County Offaly, Ireland.

When William and Laurence Parsons left England for Ireland in the latter years of the sixteenth century, it was with the intention of making their fortunes as administrative officers of the Tudor settlement of Ireland. The Parsons brothers could hardly have realized that they were founding a scientific dynasty that would, in subsequent generations, have a profound effect on science and engineering on a global scale.

Laurence Parsons was a lawyer by profession and acquired an estate of some 1100 acres at Birr in County Offaly, close to the geographical centre of Ireland. Parsons was a principal architect in the development of the town of Birr, which became known during this period as 'Parsons-town'. It is clear from historical evidence that Laurence Parsons did not share the temperament of his elder brother William, who was described by the historian W H Lecky as a rather 'unprincipled and rapacious' character. Sir Laurence, though conscientious in his duties to the Crown, seems to have managed to develop a relatively good relationship with his Catholic neighbours. Thus, while he was undoubtedly a pragmatist, he seems to have also had a reflective side to his character. It was to be these characteristics which were largely responsible for the scientific successes of his descendants.

The story now moves on to another Sir Laurence Parsons, born in 1758, who succeeded to the title of second Earl of Rosse in 1807. Parsons represented Kings County (County Offaly) in the (Irish) Houses of Parliament from 1791 to 1800 and was known as an active and vocal MP. He was described by the revolutionary Wolfe Tone as 'one of the very, very few honest men in the houses of parliament'.

The year of 1798 saw a rebellion in Ireland. The Government believed that a legislative union between Great Britain and Ireland was essential for the preservation of the security of both. The Act of Union quickly became a very contentious issue in English and Irish politics; despite fierce opposition from Irish politicians, the Act of Union became law on 1 January 1801. Laurence Parsons was very disillusioned by the imposition of the act, which effectively voted the Irish parliament out of existence. Parsons made the decision to educate his children at home and his deep personal interest in the sciences was reflected in the manner of their education. It was his eldest son, William Parsons who was to become responsible for building the great telescope at Birr, the 'Leviathan of Parsonstown'.

William Parsons was born on 17 June 1800 and was educated primarily at home until he took a first-class degree in mathematics at Magdalen College, Oxford, graduating in 1822. In 1823 he entered parliament for a short but distinguished political career, which ended in 1834 when he finally made the decision to devote his life to science. He married Mary Wilmer-Field, a wealthy heiress, on 14 July 1836. After his marriage, his parents handed over responsibility for the estate at Birr to him, and moved to Brighton, the very cold climate at Birr being uncomfortable for them. William Parsons succeeded to the title of Third Earl of Rosse when his father died in 1841.

It was his marriage to the wealthy heiress which enabled Parsons to conduct his scientific experiments, which were on such a scale as to make them hugely expensive. His wife, however, was not to be outdone; she was a most unusual woman for her time. Lady Rosse extensively remodelled the castle at Birr to her own design, and undertook such projects as the construction of a magnificent set of cast-iron gates for the castle. She found her métier, however, in the nascent science of photography, in which discipline she was awarded a silver medal by the Photographic Society of Ireland in the year 1859.

William Parsons, third Earl of Rosse died on 31 October 1867. Among his many achievements was the fact that he succeeded in passing the spirit of scientific inquiry on to his children; sadly only four of the eleven that were born survived to adulthood.

The fourth Earl of Rosse, Laurence Parsons, continued where his father had left off. After graduating from Trinity College Dublin in 1864 he went on to achieve distinction, being the first person to measure the temperature of the moon accurately. He also developed the observatory at Birr by designing automatic driving mechanisms for the instruments there. In addition to this, Laurence studied the spectra of deep-space objects in an attempt to determine their chemical composition.

Of the other sons of the third Earl, Clere Parsons became a railway engineer, and Charles Parsons invented, among many other things, the Parsons steam turbine. It was this invention which effectively allowed the industrial revolution to continue to grow, by providing a much-needed means of electrical power generation, and revolutionizing the propulsion of shipping.

To return to the central figure of our story, William Parsons, though he was audacious, was sensibly cautious in his approach to telescope construction. He built a series of telescopes ranging in size from small six-inch units right up to a 36-inch version, and along the way he honed his constructional skills, and experimented with ways of working the materials which he knew he would have to use to construct his telescope.

It is difficult to imagine now the challenges that Parsons faced when he set out to construct his huge telescope. The science and engineering required to complete it successfully just did not exist. Parsons was faced, not with a single challenge, but with a wide range of them, which would have discouraged a lesser man. The only possible approach was a very methodical one, and it is the methodical and scientific manner in which he set about solving each piece of the problem that marked Parsons out as an intellectual giant.

There were three fundamental sets of skills which Parsons needed to master in order to build the 'Leviathan'. In the first place, he needed to master the science of optics in order to address the complex issues of mirror design and construction. Mechanical engineering played a very large part as the mounting of the mirror in its tube and the accurate positioning of that tube, weighing something in the region of three tonnes, presented mechanical challenges that had never been addressed on anything like the scale that Parsons envisaged. Finally, there was a large element of materials science; the forming of optical surfaces requires an extremely demanding level of precision, which requires the use of extremely stable materials that will not change their shape by appreciable amounts over time.

The unique talent that Parsons brought to the project was his ability to demonstrate mastery in all of these areas. Additionally, when the instrument was completed, he demonstrated a very keen eye as an observer, and an excellent ability to reproduce his observations with pencil and paper. The science of photography was not sufficiently well developed at the time, and his telescope, due to the method of its mounting, was not suitable to permit him to use it for photography.

The primary mirror diameter was 72 inches and two of these were successfully completed, each mirror weighing between three and four tonnes. The mirrors were cast from speculum metal, an alloy of copper and tin. Parsons did a great deal of research into the composition of the speculum metal alloy which is a form of bronze. His final conclusion was that a ratio of 2.15 parts of copper to 1 part of tin (by weight), the same ratio that had been used by Newton, was satisfactory. This alloy, when polished, gives an optical surface reflecting about 66% of the light that is incident upon it.

Parsons also experimented with methods of mirror construction using his smaller telescope, the 36-inch instrument, as a vehicle for his research. In particular, he experimented with composite or segmented mirrors, which were cast in a series of pieces and then assembled into a

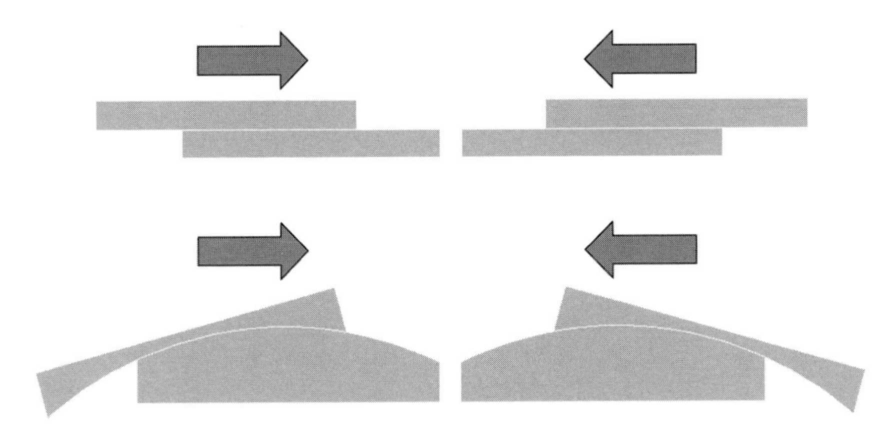

Grinding a mirror.

single unit. Though he abandoned this technique in favour of single-piece mirrors, it is interesting to note that many large-size modern mirrors have reverted to using segmented construction as the single-piece units are simply too difficult to handle.

Speculum metal is a eutectic alloy, that is to say, while copper melts at 1100°C, once it is alloyed with tin the resulting material melts at the much lower temperature of 750°C. Parsons achieved these temperatures using peat (turf as it is called in Ireland) that was in plentiful supply in the region, and he built a foundry in the moat of the castle at Birr to facilitate his experiments.

The initial melt of the components of the alloy was only the first stage in the process. The alloy was then cast into a disc using a specially constructed iron mould which was designed to have a slightly porous base to prevent gas bubbles from forming on what would eventually be the optical surface.

When the mirror blank had been cast, it was then allowed to cool very slowly over a period of some weeks. This process, called 'annealing', relieves the internal stresses in the material, and makes it easier to work. This stage is critically important, as any internal stresses will cause the mirror to change its shape over time, and the optical surface will be distorted as a result. It was at this stage that Parsons had a major setback. For his first attempt, he had insufficient tin content, and the mirror was brittle. As a result it fractured during the grinding process. Undiscouraged, Parsons analysed and explained the failure and simply set about making another one.

In total, he made five attempts and two of these, the second and fifth, resulted in usable mirrors. When one considers that each attempt required four tonnes of alloy to be melted and 2000 cubic feet of turf, not to mention three months of solid work for each attempt, one gets some idea of the scale of the project.

The cooled, annealed mirror blank had a weight of about 3.5 tonnes and a roughly concave surface. It was then ground roughly to shape, and finally polished on a steam-driven grinding and polishing machine which had been built for the purpose. The principle of operation of this machine is interesting in itself.

The top half of the illustration represents two flat discs at the beginning of the process; the bottom half shows what happens after they have been ground together for some time. The centre of the top one becomes concave and a corresponding convex surface is generated on the lower. If both pieces of material are kept in constant contact over the entirety of the stroke, there is only one shape that can result, a perfectly spherical surface. The reason this happens is that the centre of the upper disc, being in constant contact with the lower, is ground more than the edges, which overhang the lower disc for a portion of each stroke and are not ground during this portion.

Parsons built a steam-driven grinding and polishing machine to perform this operation on his mirror blanks. Once the surfaces had been ground and polished to a perfect sphere, the mirrors were then polished on a specially shaped pitch lap which converted this spherical shape to a parabolic one.

At this juncture it would be reasonable to wonder just how exacting the process of grinding and polishing the mirror actually is. As Parsons had no means of measuring the quality of his mirrors other than trying them out, no numerical data are available. It is noted that the weight of a hand on the reverse side of the three tonne mirror was all that was required to distort the image visibly. For the mirror to work well, it would probably have had to be polished to an accuracy of plus or minus half a wavelength of visible light. This is approximately 250 nm, less than one five-hundredth of the thickness of a human hair.

The extraordinary sensitivity to deformation of the mirror meant that a sophisticated support system would have to be designed for it. Parsons achieved this by floating the mirror on a complex system of supports which equalized the pressure across the back surface of the mirror, and which automatically compensated for pressure variations across the back surface as the tube was elevated.

Parsons quickly realized that the system which he would design to mount the optics would need to be massively constructed, and that a conventional, equatorial mounting of the tube would be impractical. As a result, he decided on the mechanically much simpler alt-azimuth type of mounting, which elevates the tube and allows it to swivel from side to side. This enables the observer to see a limited portion of the sky. The minimum elevation was 15°, with a maximum of 105°.

The 'Leviathan of Parsonstown'. Photograph by Mary Rosse, c 1862.

The telescope tube was a massive construction project in itself. It was constructed from pine using methods which owed much to the cooperage trade. The tube itself was 14 metres in length, with a large box at the lower end containing the mirror. The box and tube assembly was mounted on a massive cast-iron universal joint that allowed the tube to be elevated and swivelled from side to side. The entire assembly weighed something in the region of 12 tonnes.

In order to elevate and track the telescope tube, windlasses were employed. These would have been insufficient to lift the tube from its resting position of about 15° above the horizontal through the zenith and over to the other side. To alleviate this problem, a system of counter-weights was provided which kept the assembly close to perfect balance at all times. This was a highly complex task as the characteristics of the system change dramatically as the tube moves up from its resting position. Parsons' solutions to this mechanical engineering problem were brilliant. By using counterweights whose influence changed automatically with the elevation of the tube he managed to balance the system so that eight men plus the observer could operate it.

In order to give the observer some element of control, fine adjustments for azimuth and elevation were provided at the observing position. These enabled the observer to make slight corrections to the positioning of the tube, while the people at the windlasses on the ground performed the larger adjustments.

The whole telescope was supported between two massive gothic arched and castellated walls 70 feet in length and 50 feet in height.

The Leviathan was first used on 15 February 1845. Ireland was in the grip of a serious famine at the time, and the Parsons family was busy with famine relief work. The telescope was not regularly used until 1848 when it immediately proved its worth by bringing into proper view for the first time such deep-space objects as the Whirlpool Nebula M51 and the Crab Nebula M1. As well as these deep-space observations, planetary observations were made which confirmed the suspected existence of small objects in orbit around the planet Mars.

Its most significant discovery was the spiral nature of galaxies such as the whirlpool, which is very similar to our own Milky Way. This single discovery had a profound effect on our understanding of the universe in which we live. Lord Rosse also postulated, correctly, that some of the very elongated oval or 'lenticular' shaped object might be spiral nebulae whose orientation was such that they were being viewed on their sides.

The two Earls of Rosse who used the great telescope were both humble men of science; they published all of their research in detail, to enable others to follow where they had led. They made the telescope freely available to those astronomers who desired to use it, financing the post of an astronomer in residence at Birr Castle for many years.

There were many visiting scientists at Birr who went on become famous in the astronomy and other fields. Thomas Romney Robinson, who was appointed director of Armagh Observatory in 1823, played a key role in the design of the mirror support structure for the Leviathan. The Stoney brothers, George Johnstone* and Bindon Blood were also active at the observatory, lending their expertise in mathematical physics and engineering. Sir Robert Ball, later to become the Director of Dunsink Observatory in Dublin also 'cut his teeth' at the observatory in Birr which was fertile soil for his intellectual efforts.

Research with the Leviathan continued up until the death of the fourth Earl of Rosse, the son of its builder. It was dismantled in 1908 and in 1912 its remaining mirror was transported to the Science Museum in Kensington in London where it is currently on display with its support system and associated optics. It is to be hoped that eventually a means of repatriating the original mirror and optics for display at Ireland's Historic Science Centre in Birr will be found.

The remains of the telescope fell into disrepair. In 1998, the Birr Scientific and Heritage Foundation began the restoration of the great telescope at Birr Castle. This restoration is now complete, a revolutionary new optical system has been installed, and the gigantic 'Leviathan of Parsonstown' can once again peer into the deepest regions of space.

Further Reading

A History of Ireland in the Eighteenth Century, W E H Lecky (Green and Co, London, 1908).
Whatever Shines Should be Observed, S M P McKenna-Lawlor (Samton, Blackrock, 1998).
The Astronomy of Birr Castle, P Moore (Quacks Books, York, 1991).
From Galaxies to Turbines : Science, Technology and the Parsons family, W G Scaife (Institute of Physics, Bristol, 2000).

Humphrey Lloyd

1800–1881

James G O'Hara

The experimental physicist Humphrey Lloyd, eldest son of the Rev Bartholomew Lloyd (1772–1837) and Eleanor McLaughlin, was born in Dublin on 16 April 1800. He was a direct descendant of Humphrey Lloyd (1656–1727) who came from Denbighshire in Wales and settled in County Wexford in 1683; the family tree reveals the names Bartholomew and Humphrey alternating as direct ancestors over five generations. He received his early education at a private school, Mr White's School, Dublin, before entering Trinity College in 1815, gaining first prize among 63 competitors at the entrance examination. He profited greatly from the renaissance in mathematical education, initiated at the College by his father, Bartholomew Lloyd, who was Professor of Mathematics, Natural Philosophy, Greek and Divinity, and in 1831 became Provost of the College.

Following a brilliant undergraduate career during which he obtained a scholarship in 1818, Humphrey Lloyd graduated BA in 1819, taking first place and the gold medal for science; he became a Fellow of the College in 1824 and received an MA in 1827. Thus embarked on an academic career, he commenced studies and research in physical science, and in 1831 he succeeded his father as Professor of Natural and Experimental Philosophy. During his tenure of that chair he endeavoured with considerable success to upgrade the state of physics and physics teaching within the university.

In physical optics he made a number of noteworthy contributions. Most remarkable was the experimental proof, in December 1832 and January 1833, of the existence of conical refraction in a crystal of arragonite, following a prediction arising from a mathematical investigation of the Fresnel wave surface in biaxial crystals by his colleague William Rowan Hamilton* (see the articles on Hamilton and also James MacCullagh*). Lloyd demonstrated two species of conical refraction, for light entering and leaving such a crystal, and he established the law of conical polarization, the law governing the polarization of the rays in the refracted cone.

Conical refraction was, at first sight, no more than a curious optical phenomenon and offered no conceivable application, its significance

Humphrey Lloyd (courtesy of the Royal Society, London).

being entirely theoretical. It represented a vindication of Fresnel's theory of double refraction, and was the final chapter in the history of the theory of double refraction, which had begun a century and a half earlier with the first formulation of the wave theory by Christiaan Huygens to explain the phenomenon. It was a further triumph for the wave theory of light over the particle theory and was a cause of excitement among the supporters of the wave theory. In the history of physics, it is an example of a mathematical prediction being verified by experiment, in contrast to the more common case of observed physical phenomena being subsequently explained by mathematics.

Of significance too was Lloyd's preparation of a substantial report on *The Progress and Present State of Physical Optics* (published in the *British Association Report*, 1834) which was widely acclaimed. Another success in the area of physical optics followed shortly afterwards, when he demonstrated the interference of light passing directly from a luminous source, with that coming from the same source but reflected at an angle of incidence of nearly 90° from a plane surface. Prior to Lloyd, Augustin Fresnel had demonstrated interference fringes from rays of light from a single source that are reflected by two plane mirrors that subtend an angle of almost 180°. Lloyd's single mirror experiment demonstrated two important properties of reflected light, namely the intensity of the light reflected at nearly 90° incidence is equal to that of the direct light, and that a half-wavelength phase change takes place on reflection at the interface of the higher density medium. The conditions of Lloyd's experiment are very similar to those governing the passage of light down the core of a monomode optical fibre in today's optical networks.

However, Lloyd's variation of Fresnel's famous twin-mirror experiment, which he published in 1837, was—like the discovery of conical refraction—regarded as important not for any conceivable practical applications but rather as further proof of the correctness of the wave theory of light, of which he was an ardent supporter. At a British Association for the Advancement of Science annual meeting at Manchester in 1842, where a vigorous debate took place between protagonists and antagonists of the wave theory, he was in the vanguard of the assault on the particle theory. Lloyd also investigated the phenomena of light incident on thin plates and in 1841 he submitted a communication on the subject to the British Association. In 1859 he presented his complete investigation of the phenomenon to the Royal Irish Academy.

Investigation of the earth's magnetic field was Lloyd's primary research interest from the mid 1830s. Working at first under the aegis of the British Association, he achieved prominence in the field of instrumentation. Having devised a method for measuring dip and relative intensity with the same instrument, he carried out, in collaboration with the Dublin-born Colonel Edward Sabine and James Clerk Ross, a magnetic survey of Ireland in 1834 and 1835. In the two following years this work was extended to Scotland, and then to England and Wales.

At the end of 1835 Lloyd established contact with Carl Friedrich Gauss of Göttingen who, in developing the theory and technique of absolute measurement of the earth's field, had established geomagnetic research on a new foundation. Before Gauss, only relative or comparative measurement of the earth's magnetic field had been possible. Gauss's contribution, published in a work entitled *Intensitas Vis Magneticae Terrestris ad Mensuram Absolutam Revocata* (On the Intensity of the Terrestrial Magnetic Force in Absolute Measure) in 1833, was important for two main reasons. Firstly, the terrestrial magnetic force was for the first time properly expressed in absolute units, and secondly, a method was introduced for the measurement of the horizontal component at a given place in absolute units, and independently of the magnetic properties of the bar or needle used.

In the ten years following its invention, Gauss's method was extensively applied in geomagnetic research, which had become an exciting research field, and in which extensive international cooperation had developed. Although mathematically and physically sound, Gauss's

The Magnetic Observatory of Trinity College Dublin (from a drawing made in 1842).

method presented some unforeseen problems in its practical implementation. The solution of these problems required the ingenuity and experience of a number of younger physicists working in observational and experimental physics. Lloyd was a pioneer in the development of observational techniques based on Gaussian principles in this period.

Following the magnetic surveys of the British Isles, Lloyd resolved to join the organization of observing stations established by the Hanoverian mathematician over the northern hemisphere. An observatory was built at Trinity College Dublin in 1837–1838, and fitted out with instruments of Lloyd's design operating on Gaussian principles and constructed for the most part by the Dublin instrument maker Thomas Grubb*. To measure the terrestrial magnetic field at a specific location it was necessary to measure the declination, the inclination or dip, and either the horizontal or vertical component of the earth's magnetic field (or alternatively both of the components and one of the angles). Besides making occasional determinations of the absolute value of the field at a geographical location, it was intended to monitor the changes taking place in the field, by a regular series of observations continued over an extended period of time. Furthermore the behaviour of the field around the globe was to be observed and charted at fixed stations as well as by travellers.

Lloyd devised six principal instruments for his observatory, namely a declinometer, an inclinometer or dip circle, a unifilar or single-thread suspension magnetometer, a bifilar or twin-thread suspension magnetometer, a balance magnetometer and an induction inclinometer. The latter three were novel special-purpose instruments used for observing the variation of the terrestrial magnetic field. The instruments designed by Lloyd were characterized by a high degree of accuracy. For example, the instruments for measuring the changes in the two components of the intensity of the field could read within 1/50 000th of the whole.

Lloyd's magnetical observatory became the model for a series of similar observatories in Britain and the colonies established, following a

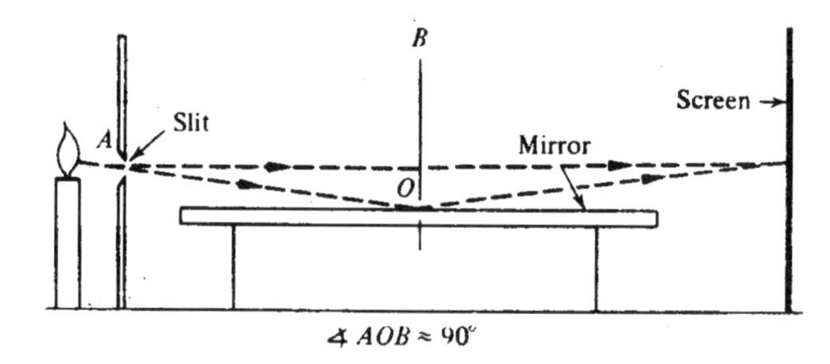

Lloyd's mirror experiment.

recommendation of the Royal Society, by the government and the East India Company in 1839. These, like the Antarctic expedition (1839–1843) led by Ross, were provided with instruments similar to those of Lloyd, and the observers received practical instruction from him at his Dublin magnetical observatory. The worldwide network of observing stations superseded Gauss's earlier organization, and many continental stations joined the association. In all 33 observatories were established or re-equipped with instruments of Lloyd's or a compatible design, and the work was continued for nine years (1839–1848). In connection with Arctic expeditions in 1845 and in 1848, Lloyd designed an instrument to be used by travelling observers and by mariners.

The excitement at the extent of the scheme is captured in the following extract from Lloyd's *Account of the Magnetical Observatory of Dublin* published in 1842: "The distinguishing characteristic of this undertaking,—that which gives it unity and greatness,—is, that the same plan of observation is followed out in all these distant stations, by observations strictly simultaneous, made according to the same instrumental methods, and with the same instrumental means. In order to give a still wider extension to the scheme, which was matured under its direction, the Royal Society solicited the cooperation of foreign states. The invitation was responded to in a spirit unparalleled in the history of science. Most of the foreign observatories were reorganized, on a scale of greater completeness; many new ones were added; and thirty-three observatories are now in operation, following out the same plan in all its details.

"Of these eleven (including Greenwich and Dublin) are established in Britain and her dependencies; and ten have been founded and equipped by the Russian Government, viz., at Petersburg, Catherineburg, and Kazan, in Russia proper; at Helsingfors, in Finland; at Nicolajeff, in the Crimea; at Tiflis, in Georgia; at Barnaoul and Nertchinsk, in Siberia; at Sitka, in North America, and at Pekin in China. Of the rest, one has been established by the French Government at Algiers; one by the Belgian, at Brussels; two by Austria, at Prague and Milan; one by Prussia at Breslau; one by the Bavarian Government at Munich; and one by the

Spanish, at Cadiz; there are two in the United States, at Philadelphia and Cambridge; one at Cairo, founded by the Pasha of Egypt; one at Trevandrum, in India, by the Rajah of Travancore; and one by the King of Oude, at Lucknow. The observatories at Brussels, Breslau, Cadiz, Cambridge, Algiers, Cairo, Trevandrum, and Lucknow, are provided with instruments similar to those of Dublin."

A range of papers Lloyd wrote on geomagnetism and other related subjects, such as meteorology, were published in the *Reports* of the British Association, and in the journals of the Royal Irish Academy. He was President of the latter body from 1846 to 1851, and in 1862 the Academy awarded him its Cunningham gold medal. By then Lloyd had reached the zenith of his academic career. In 1843 he had been made a Senior Fellow of Trinity College; in 1862 he became Vice-Provost and, in 1867, Provost. He received a series of awards and distinctions outside Ireland; he became a Fellow of the Royal Societies of London and Edinburgh, and an honorary member of a series of other learned societies of Europe and America. In 1855 the University of Oxford conferred on him the degree of DCL, and in 1857 he was president of the British Association when it met in Dublin.

Lloyd journeyed on at least four occasions to the continent. Accompanied by Edward Sabine, he visited Berlin, Leipzig and Göttingen in the autumn of 1839. He met important scientists such as Gauss, Wilhelm Weber, and Alexander von Humboldt, pioneer of geophysical investigation and patriarch of German science. Following his marriage in the second week of July 1840, Lloyd and his wife travelled through Switzerland, Northern Italy, Tyrol and Bavaria. They visited the observatories at Milan and Munich, and joined the meeting of the Italian Physical Society at Turin in September, before reaching the Brussels observatory around 24 October. In the summer of 1841 he travelled with his wife to Paris to study the French system of training engineers, and on a further tour with his wife he visited Berlin in early October 1849 as President of the Royal Irish Academy, and was received there by Alexander von Humboldt. Probably through Humboldt's influence the Emperor of Germany awarded him the order of merit, 'Pour le Mérite', in 1874.

Lloyd no doubt shared the pro-Union political outlook of most members of the Irish Anglican ascendency in the mid and late nineteenth century. In university administration, educational and ecclesiastical policy, he was a man of firm principle, but one who was open for reform and innovation. He contributed to the abolition of religious tests in Trinity College Dublin in 1873, and he was a leading member of the general synod of the disestablished Church of Ireland after 1870, writing works on doctrinal issues, and contributing to the debate on the revision of the Church prayer book. His religious outlook was close to that of the evangelical, anti-sacerdotal wing of the Church.

Humphrey Lloyd, like his father, may be regarded as a progressive educator of the nineteenth century. His scientific philosophy was firmly Baconian and his commitment to the scientific method and the central role of experiment was already evident in his *Two Introductory Lectures on Physical and Mechanical Science* of 1834. He was primarily responsible, in 1850, for the introduction of the first Science Moderatorship (equivalent to the celebrated Cambridge Tripos examination) in Experimental Physics, and was the founder of a tradition of scientific research, particularly in physics, within the University of Dublin. He contributed to the advancement of science and engineering education in the University and was instrumental in the establishment of the School of Engineering that opened in November 1841, helping to create Chairs of Geology and Mineralogy, and of Applied Chemistry.

His mature philosophy of education is revealed in a pamphlet *Brief Suggestions in Reference to the Undergraduate Curriculum in Trinity College*, published anonymously in 1869, in which he stressed the importance of mental training, of preparation for professional life, and of a broad and varied curriculum, combined with a freedom of choice for advanced students. He advocated that English language and literature and the sciences should occupy a more prominent place, at the expense of mathematics and classics, which had previously dominated undergraduate studies.

Until 1840 the junior fellows of Trinity College were required to be celibate and to take holy orders in the Church of Ireland. Following abolition of the celibacy rule for fellows, Lloyd married, in July 1840, Dorothea Bulwer, daughter of a Norfolk clergyman. They had no children, and accordingly he was the last direct descendant of Humphrey Lloyd (1656–1727). Lloyd spent his entire life in Ireland, residing at various times at Trinity College, at 17 Fitzwilliam Square South, at Killcroney Abbey, Enniskerry, County Wicklow, and at Victoria (now Ayesha) Castle, Killiney, County Dublin. He died at Provost's House, Trinity College Dublin on 17 January 1881.

An attractive bust of him by Albert Bruce Joy was donated by his widow and stands in the Long Room of Trinity College Dublin. Two perhaps less impressive portraits also belong to the College. One of these, painted from a photograph, was presented in 1916 to the College by Dean C J Ovenden of St Patrick's Cathedral; the other portrait is probably a copy of that by Dean Ovenden. The Royal Irish Academy owns a death mask which is thought to be that of Lloyd. The best likeness of Lloyd in his prime may be a portrait drawing, with his signature, made at Brussels in 1840, that is in possession of the Royal Society of London. The Royal Society also possesses a photograph of him as an elderly man. Perhaps the most appropriate and visible monument to Humphrey Lloyd is his magnetical observatory, which stood in the Provost's garden at Trinity College from the time of its erection in 1838 until 1974 when it was removed; the façade now stands on the campus of University College Dublin at Belfield.

Humphrey Lloyd's publications are the fruits of a career in research and teaching spanning almost half a century. He published, in addition to a circular for directors of magnetic observatories, and university addresses and lectures, a total of eight textbooks or monographs and 64 papers (three jointly with others) on scientific topics (including reports to the British Association). A German translation of his *Report on the Progress and Present State of Physical Optics* (*British Association Report* for 1834) was published in Berlin (1836).

His more important scientific publications include the following: *A Treatise on Light and Vision* (1831), *Two Introductory Lectures on Physical and Mechanical Science* (1834), *Lectures on the Wave Theory of Light*, published in two parts (1836 and 1841) and later republished as an *Elementary Treatise on the Wave Theory of Light* (1857), *Account of the Magnetical Observatory of Dublin, and of the Instruments and Methods of Observation Employed there* (1842), *An Account of the Method of Determining the Total Intensity of the Earth's Magnetic Force in Absolute Measure in High Magnetic Latitudes* (1848), *The Elements of Optics* (1849), *Observations Made at the Magnetical and Meteorological Observatory at Trinity College, Dublin* (1865), *Treatise on Magnetism, General and Terrestrial* (1874), and *Miscellaneous Papers connected with Physical Science* (1877). These works reflect the outstanding career of an Irish physicist who in his day enjoyed considerable international renown.

Further Reading

Dictionary of National Biography and the forthcoming *New Dictionary of National Biography*.

'Humphrey Lloyd 1800–1881', T D Spearman, *Hermathena* **130** (1981) 37–52.

'The Prediction and Discovery of Conical Refraction by William Rowan Hamilton and Humphrey Lloyd (1832–1833)', J G O'Hara, *Proceedings of the Royal Irish Academy* **82A** (1982) 231–257.

'Gauss and the Royal Society: The Reception of his Ideas on Magnetism in Britain (1832–1842)', J G O'Hara, *Notes and Records of the Royal Society of London* **38** (1983) 17–78.

'Gauss' Method for Measuring the Terrestrial Magnetic Force in Absolute Measure: Its Invention and Introduction in Geomagnetic Research', J G O'Hara, *Centaurus* **27** (1984) 121–147.

Optics after Newton: Theories of Light in Britain and Ireland 1704–1840 G Cantor (Manchester University Press, 1983).

Thomas Grubb and Howard Grubb

1800–1878 1844–1931

Andrew Whitaker

The manufacturing firm headed first by Thomas Grubb, and later by his son Howard, remained in business for practically a century from 1830. Reformed as Grubb–Parsons in 1925, it continued for another 60 years. The firm was unique in nineteenth-century Ireland as an advanced technological enterprise. It had quite a wide range of activities. For much of his life, the majority of Thomas's efforts were directed towards, and most of his income was derived from, his appointment as Engineer to the Bank of Ireland; in this capacity his ingenious inventions were responsible for substantial improvements in the engraving, printing and numbering of banknotes. He also patented a design for a cheap camera lens which sold well. From the turn of the century, as nations prepared for war, and submarines were rapidly constructed, much of the effort of the firm went into construction of periscopes and other equipment for the military.

Only a part of the effort of the Grubbs went into telescope design and construction, but it was this activity that has led to their continuing fame, and it may be assumed that it represented the central interest of both father and son. The firm was responsible for some of the world's largest and best telescopes. As well as displaying great talent in developing the optics of these devices, the Grubbs were able to achieve substantial improvements in the mechanical aspects of telescope design, the stability, and the capacity to control reliably the position of the telescope axis, and to vary it systematically for considerable periods in order to track stars.

Thomas Grubb was born in Waterford on 4 August 1800. Little is known of his early life. His rather primitive writing style in later life may suggest that he had comparatively little formal education, though he was undoubtedly an original thinker, a considerable inventor, and a natural solver of any type of problem. In the 1820s he was a clerk in Dublin, but by the early 1830s he had an engineering business near Charlemont Bridge in Dublin. His products included machine tools and cast iron beds for billiard tables. However, it is clear that his interest in astronomy was already developed. His premises included an observatory, substantial enough that members of the public were invited to

Thomas Grubb (1800–1878) (Mary Lea Shane Archives, Lick Laboratory).

pay to view it, and he manufactured telescopes and other astronomical instruments as well as more mundane items.

It seems clear that Grubb's interest in astronomy was stimulated by (Thomas) Romney Robinson, a noted mathematician, physicist and astronomer, who was Director of the Armagh Observatory from 1823 until his death in 1882, though it is not clear how their acquaintanceship arose. There were to be many connections between the Grubb and Robinson families. One of Thomas Grubb's daughters, Mary Ann, married Romney Rambaut, Robinson's nephew. Robinson's daughter married G G Stokes*, the Cambridge mathematician and physicist, who, as a long-term consultant to the Grubb firm, provided inestimable support on advanced optical theory. Robinson was also well acquainted with William Parsons, later Earl of Rosse*, who was to become famous as the builder of telescopes, and this was also to be an extremely important contact for Grubb; both parties profited from exchange of information and ideas.

Thomas Grubb's first step towards world renown as a manufacturer of telescopes came in the early 1830s with his construction of an equatorial mounting for the 13.3-inch lens of Markree Observatory. Markree was the County Sligo estate of Edward Cooper, who had decided to develop an observatory at the site. He purchased the lens, which had a focal length

Howard Grubb, probably aged about 35 (Mary Lea Shane Archives, Lick Laboratory).

of over 25 feet, from R A Cauchoix of Paris, but originally it was supported by a crude alt-azimuth stand, and the results obtained from this giant instrument, which had the largest aperture in the world at the time, were disappointing.

Romney Robinson had become acquainted with Cooper, and he pointed out that the excellent capability of the lens was effectively being wasted by the mechanical aspects of the system, by poor centring and by the lack of stability and flexibility of the support system. Robinson persuaded Cooper to order an equatorial mounting from Grubb.

Grubb's work at Markree marked a major advance in ensuring the stability and control of massive telescopes. The telescope was erected on a triangular pier of black marble blocks. The axes were large solid castings which rotated within strong bearings. The driving clock was extremely sophisticated for its period; it was regulated by a governor which increased frictional drag if the clock was going fast and reduced it if was going slow. The gearing between the clock and the telescope controls was designed to the highest standards. As with all Grubb's work, the quality of the engineering was superb. This was in the period before the invention of ball bearings, and Grubb used many clever schemes to reduce friction between moving parts, while maintaining the stability of the telescope axis.

Robinson now asked Grubb to provide a mounting for a 10-inch mirror for Armagh Observatory, though, as events turned out, by the time the instrument was produced in 1835, it had become a 15-inch telescope, and Grubb had provided a new speculum metal mirror, as well as a driving clock. This was the first reflecting telescope of any size to have been given a respectable equatorial mount. (Until then, the fact that mirrors were much cheaper than lenses encouraged operators to economize also on their mounting.)

It was also the first time that a system of balanced levers moving on simple bearings was used to support the mirror, rather than resting and moving the telescope on small wheels. This enabled a heavy mirror to retain its shape irrespective of its orientation with respect to the vertical. These advances in design made possible a considerable improvement in performance over previous reflecting telescopes. They were to be used by the Earl of Rosse and many others.

Over the next 20 years or so, Grubb designed several more telescope systems: he performed the mounting for the Sheepshanks refractor which had been presented to the Royal Greenwich Observatory; he sold to the United States Military Academy at West Point a substantial telescopic system of 15-inch aperture, which was to be used to make the first quantitative astronomical measurements by photography; he reworked a reflecting telescope of 22-inch aperture for Glasgow University Observatory; and he constructed a refractor for Dunsink Observatory. For the last Grubb built a rotating dome, a feature which was to become a speciality of the firm.

However, his career as a man of science also developed in a much richer vein. In 1835 he became a member of the British Association for the Advancement of Science. In 1836 he read a paper on means of improving reflecting telescopes to the Royal Astronomical Society. He constructed equipment for James MacCullough*, which MacCullough used in 1838 to demonstrate to the Royal Irish Academy experiments related to his theory of light reflection from metals. In 1839 Grubb himself became a Member of the Royal Irish Academy, a singular honour for one who had started as a simple instrument maker with little formal education.

Another important scientific development in which Grubb played a major part was that led round about 1840 by Humphrey Lloyd*, Professor of Natural Philosophy at Trinity College Dublin, which established identical magnetic stations at all parts of the globe in order to build up a full understanding of the magnetic properties of the earth. Grubb provided most of the equipment for all forty stations.

Grubb's theoretical repertoire was also expanding. By the 1850s he was one of the first to study the behaviour of optical instruments by use of ray tracing; he felt that mathematical methods on their own were too cumbersome to be useful for practical problems, but the mathematical

The casting of a speculum for the Great Melbourne Telescope. The crucible of molten speculum metal is being moved to the tipping frame, from where it will be poured into the circular mould. "Every man wore a large apron and gauntlets of thick felt, with an uncanny-looking calico hood, soaked in alum, drawn completely over his head. This hood was provided with large, glistening talc eyes. These weird figures flitted about in the ghastly light of the intense soda-flame that leapt from the great furnace, and the windows were filled with the eager faces of fascinated spectators." [From W. G. Fitzgerald, *Strand Magazine* **12** (1896) 369, quoted in *Victorian Telescope Makers*, I S Glass (Institute of Physics, Bristol, 1997)].

and practical methods together could be highly successful. With their use, he designed a wide-angle achromatic camera lens, which sold extremely well. Indeed, he had a very strong lifelong interest in photography. He also used his technique to design and construct improved microscopes. The pinnacle of this scientific achievement was his election as a Fellow of the Royal Society in 1864.

For all this scientific success it seems certain that Grubb was extremely keen to get back to building a huge telescope. In 1866 he got the chance with the projected Great Melbourne Telescope. The original plan for the operation of an extremely large reflecting telescope in the southern hemisphere dated from the 1840s, but no funds were provided by the British Government for practically quarter of a century.

When this changed, the Royal Society Committee made the fateful decision make to use speculum metal rather than metal-on-glass for the mirrors. All the members of the Committee were experienced hands-on astronomers, who perhaps did not recognize that, in the hands of the

The Great Melbourne Telescope assembled in Grubb's yard in Dublin.

inexperienced, the heavy re-polishing that speculum metal mirrors require may severely damage their optical properties.

The construction of the Melbourne Telescope was a massive task for the Grubbs. It was at this time that Thomas' son, Howard, joined the firm. Thomas and his wife Sarah had had nine children, but only four, two sons and two daughters, lived into adulthood. Of the two sons, Henry, who was born in 1833, eventually took over his father's position as Engineer to the Bank of Ireland. Howard, the last child in the family, was born in 1844. By 1866 he was studying Engineering at Trinity College Dublin, but he was then removed from the College before finishing his degree, and put in charge of the Melbourne Telescope.

At the same time, the whole enterprise was upgraded. A site at Rathmines was bought, and over the next few years a totally new Optical and Mechanical Works was constructed, with specialized workshops for all the various aspects of the work, and large assembly areas where the completed instruments would be put together. At its peak, the firm employed many skilled and semi-skilled workers—engineers, draughtsmen, instrument makers, glass grinders and opticians.

In the book listed in Further Reading, Glass gives a full account of the huge effort put into construction of the Melbourne Telescope on the Dublin premises. The speculum consisted of four parts of copper to one of tin, to reduce the rate of tarnishing. Annealing and casting were hard dangerous tasks. Then followed grinding and polishing. Finally came the mounting; as usual for Grubb the support was massive to provide stability, but the system of levers meant that there were very small forces on the bearings. The Royal Society Committee described the telescope as 'a masterpiece of engineering'.

The telescope was dismantled for shipping to Australia in April 1868, the specula being varnished for protection, and the telescope arrived in Australia in November. Sadly what followed was anti-climax. The varnish was removed badly, the speculum began to tarnish, and despite considerable efforts, there was no one in the country able to organize effective re-polishing. The instrument was never used to very great effect, and sadly the episode had the unfortunate effect of hindering the development of reflecting telescopes for the next 30 years.

At the time, though, the construction of the Melbourne Telescope made the Grubbs famous. From now on they were concerned with, or gave advice on, virtually every large telescope project in the world. The last major project of Thomas Grubb's life, though achieved largely by his son, was the Great Vienna Telescope. The Grubbs were awarded the contract for this instrument, which was to be the largest refracting telescope in the world, in 1875.

The main architectural feature of the Vienna Telescope was a great central dome, 45 feet in diameter. The total weight of this dome, which was constructed in Dublin, was around 15 tons, but it took a force of only 70 lb to rotate it. This dome was surrounded by three smaller domes. The refractor was 27-inch, though this was to hold the world record only until 1885. The instrument was working in Dublin by 1881, it left for Vienna in 1882, and by 1883 it was clear that it was a fine instrument on which excellent work had already been performed.

Thomas Grubb had suffered from rheumatism since the early 1870s, and his role in the firm had gradually decreased. He died in 1878.

The last 20 years of the century were perhaps the busiest period of Howard Grubb's career, at least as far as the building of telescopes is concerned. He made around 90 telescope objective lenses from 5 to 28 inches in diameter, and also most of the required mountings and controls. The countries where his telescopes were constructed include South Africa, India, the USA, Germany, Mexico, Australia, Spain, Venezuela, Belgium, Bulgaria, New Zealand and Turkey, and, of course, many telescopes were provided in Ireland, England and Scotland.

While Thomas Grubb had been highly original and inventive, Howard was perhaps more practically minded. Rather than seeking major conceptual advances in telescope design, he was largely content to move forward in piecemeal fashion as customers demanded improvements, and his methods tended to be based on trial-and-error rather than fundamental theory.

For all that, he was a reputable scientist, publishing over 60 papers, a large number of which were in the publications of the Royal Dublin Society. Many of them covered matters specifically to do with telescope design, but they also include other areas of optics, and topics further afield still. He was an officer of the Royal Dublin Society for many years, and received its Boyle Medal from its President, John Joly*, in 1912. He was awarded the Cunningham Medal of the Royal Irish Academy in 1881. He followed his father by being elected as a Fellow of the Royal Society in 1883, and was knighted in 1887.

One of the largest projects in which Grubb took part in these years was the Astrographic Project. In 1887 an international congress was arranged to coordinate sky mapping using photography; it recommended that a *Carte du Ciel* (Map of the Sky) should be produced, each observatory participating in the scheme mapping a particular zone of declination using standard telescopes, the aperture of which was to be 33 cm, and the focal length 3.43 m. Grubb built seven of the telescopes—those for Greenwich, Oxford, Cork, the Cape of Good Hope, Sydney, Melbourne and Tacubaya (Mexico).

From roughly the turn of the century, Grubb's work on telescopes lessened in scope. Economic power was gradually moving towards the United States, where companies with far more financial clout than the Grubbs could have dreamed of were available to build new generations of telescopes. The emphasis of the Grubb firm moved towards military projects. Grubb himself designed a novel form of optical gun-sight and was also strongly involved in the design of periscopes for submarines.

Particularly once the First World War came, this work was extremely lucrative, and during the war Grubb supplied around 95% of the periscopes for British submarines. (Before the war, the Germans had also bought some of Grubb's periscopes, so the U-boats may have been using the same model as their enemies.) The factory also made large quantities of gun-sights and rangefinders.

However, the British Admiralty were not prepared for this work to continue in Ireland, because of danger and delays in transport, the political unrest in Ireland, and the desire to have the work carried out near to headquarters for consultation and inspection. The work was transferred to St Albans in Hertfordshire, England, where a new factory was provided for Grubb. However, the move was only partially completed when the war ended.

At this time Grubb wished to get back into telescope work; he had a pre-war contract from the Imperial Russian Government to build several telescopes and an observatory in Russia, a contract which the Soviet Government honoured, and also a contract for a 26.5-inch refractor for Johannesburg. However, the prices for these telescopes had not risen since before the war. Wartime inflation and the high wages Grubb found he needed to pay for skilled manpower meant that the company could not survive. It went into liquidation in 1925.

There was to be a rescue. Sir Charles Parsons, youngest of the six sons of the third Earl of Rosse, purchased the firm. He was himself well-known for the invention and utilization of the steam turbine, but also had a good deal of experience in optical matters. The new firm of Grubb–Parsons had its headquarters in Newcastle-upon-Tyne, and continued to make major telescopes until 1985.

Howard Grubb returned to Ireland, living first at Dun Laoghaire, then later at Monkstown. He died on 16 September 1931.

Further Reading

Victorian Telescope Makers: The Lives and Letters of Thomas and Howard Grubb, Ian S Glass (Institute of Physics, Bristol, 1997). This is a standard work providing a wealth of information on the Grubbs and all their projects. Its use in the preparation of this article is fully acknowledged.

Howard Grubb: Inventor and Manufacturer of Scientific Instruments, Gordon Herries Davies in: *Some People and Places in Irish Science and Technology*, Charles Mollan, William Davis and Brendan Finucane (eds) (Royal Irish Academy, Dublin, 1985), pp 60–61.

Thomas Grubb: Engineer and Optician, Patrick A Wayman in: *More People and Places in Irish Science and Technology*, Charles Mollan, William Davis and Brendan Finucane (eds) (Royal Irish Academy, Dublin, 1990), pp 18–19.

Articles by H C King on Howard Grubb and Thomas Grubb in *Dictionary of Scientific Biography*, C C Gillespie (ed) (Scribner, New York, 1970–80).

The History of the Telescope, H C King (Griffin, High Wycombe, 1955; reissued by Dover, New York, 1979).

William Rowan Hamilton

1805–1865

John J O'Connor and Edmund F Robertson

William Rowan Hamilton was born in Dublin at midnight on 3–4 August 1805. His father, Archibald Hamilton, was often away in England pursuing legal business. He had not had a University education and it is thought that Hamilton's genius came from his mother, Sarah Hutton. Because of his absences, Archibald was not able to teach William, but this may have been an advantage for the boy. Instead he was taught by his uncle, the Rev James Hamilton, with whom William lived in Trim for many years. James was a fine teacher, and, by the age of five, William had already learned Latin, Greek and Hebrew.

William soon mastered additional languages, but a turning point came in his life at the age of 12 when he met the American Zerah Colburn. Colburn could perform amazing mental arithmetical feats, and Hamilton joined in competitions of arithmetical ability with him. It appears that losing to Colburn sparked Hamilton's interest in mathematics.

Hamilton's formal introduction to mathematics came at the age of 13, when he studied Clairaut's *Algebra*, a task made somewhat easier as Hamilton was fluent in French by this time. At age 15 he started studying the works of Newton and Laplace. In 1822 Hamilton found an error in Laplace's *Méchanique Céleste* and as a result of this came to the attention of John Brinkley, the Astronomer Royal of Ireland, who said: 'This young man, I do not say will be, but is, the first mathematician of his age'.

Hamilton entered Trinity College, Dublin at the age of 18, and in his first year he obtained an *optime* in Classics, a distinction only awarded once in 20 years. He achieved this merit despite spending most of his time living with his cousin Arthur at Trim and therefore not attending all of his lectures.

In August 1824, Uncle James took Hamilton to Summerhill to meet the Disney family. It was at this point that William first met their daughter Catherine, and immediately fell hopelessly in love with her. Unfortunately, as he had three years left at Trinity College, Hamilton was not in a position to propose marriage. However, Hamilton was making remarkable progress for an undergraduate, and submitted his first paper entitled *On Caustics* to the Royal Irish Academy before the end of 1824.

William Rowan Hamilton (courtesy of the Royal Irish Academy).

The following February, Catherine's mother informed William that her daughter was to marry a clergyman 15 years her senior. He was affluent, and could offer more to Catherine than Hamilton could. In his next set of exams William was so distraught at losing Catherine that he was given only a *bene* instead of his usual *valde bene*. He became ill and at one point he even considered suicide. It was at this period that he turned to poetry; he was to do the same for the rest of his life when he was in anguish.

In 1826 Hamilton received an *optime* in both science and classics, a feat which was unheard of, and in his final year as an undergraduate he presented a memoir *Theory of Systems of Rays* to the Royal Irish Academy. It is in this paper that Hamilton introduced the characteristic function for optics. This was the start of his most lasting contribution to mathematical physics.

Hamilton's finals examiner, Boyton, persuaded him to apply for the post of Astronomer Royal at Dunsink observatory, even although there had already been six applicants, one of whom was George Biddell Airy, who was to become Astronomer Royal less than ten years later. Later in

1827 the Board of Trinity College appointed Hamilton Professor of Astronomy while he was still an undergraduate aged 21 years. This appointment brought a great deal of controversy, as Hamilton did not have much experience in observing. His predecessor, Professor Brinkley, who had become a bishop, did not think that Hamilton had been wise to accept the post, and implied that it would have been prudent for him to have waited for a fellowship. It turned out that Hamilton had indeed made a poor choice; he lost interest in astronomy and spent all his time on mathematics.

Before beginning his duties in this prestigious position, Hamilton toured England and Scotland, from where the Hamilton family originated. During this trip he met the poet Wordsworth and they became friends. One of Hamilton's sisters Eliza wrote poetry too, and later, when Wordsworth visited Dunsink, it was her poems that he liked rather than Hamilton's. The two men had long debates over science versus poetry. Hamilton liked to compare the two, suggesting that mathematical language was as artistic as poetry. However, Wordsworth disagreed saying that: "Science applied only to material uses of life waged war with and wished to extinguish imagination" (as quoted in the biography by Hankins in Further Reading).

Wordsworth had to tell Hamilton quite forcibly that his talents were in science rather than poetry: "You send me showers of verses which I receive with much pleasure... yet have we fears that this employment may seduce you from the path of science.... Again I do venture to submit to your consideration, whether the poetical parts of your nature would not find a field more favourable to their nature in the regions of prose, not because those regions are humbler, but because they may be gracefully and profitably trod, with footsteps less careful and in measures less elaborate."

It was at about this time that Hamilton took on a pupil by the name of Adare. Unfortunately Adare's eyesight started to present problems as he was doing too much observing, while at the same time Hamilton became ill due to overwork. To recover they decided to take a holiday in Armagh where they visited another astronomer, Romney Robinson. It was on this occasion that Hamilton met Lady Campbell, who was to become one of his favourite confidantes. He also took the opportunity to visit Catherine who was living quite close to Armagh. In turn she came to the Observatory. Hamilton was so nervous in her presence that he broke the eyepiece of the telescope whilst trying to give her a demonstration. This episode inspired another interval of misery and the writing of poetry.

In July 1830 Hamilton and his sister Eliza visited Wordsworth, and it was around this time that he started to think seriously about getting married. He considered Ellen de Vere, telling Wordsworth that he 'admired her mind', but he did not mention love. He did, however, bombard her with poetry and was about to propose marriage when, according to the biography by O'Donnell in Further Reading, she happened to say that she 'could not live happily anywhere but at Curragh'. Hamilton thought this was her way of discouraging him tactfully and so he ceased to pursue her. However, he was shown to be mistaken, as she did marry the following year, and did leave Curragh! Fortunately, one good thing transpired from the event, as Hamilton became firm friends with Ellen's brother Aubrey, although, following a dispute about religion in 1851, they went their separate ways.

Catherine aside, Hamilton seemed quite fickle when it came to relationships with women. Perhaps this was because, if he could not have Catherine, then it did not really matter to him who he might marry. Eventually he married Helen Maria Bayly, who lived just across the fields from the Observatory. William told Aubrey that she was 'not at all brilliant', and unfortunately the marriage was fated from the start. They spent their honeymoon at Bayly Farm; Hamilton spent the whole time working on the third supplement to his *Theory of Systems of Rays*. Then at the Observatory Helen did not have much of an idea of housekeeping, and was so often ill that the household became extremely disorganized. In the years to come she was to spend most of her time away from the Observatory, as she was either looking after her ailing mother or was indisposed herself.

The third supplement mentioned above was published in 1832. It was essentially a treatise on the characteristic function applied to optics. Near the end of the work he applied the characteristic function to study Fresnel's wave surface; from this he predicted conical refraction, and asked the Professor of Physics at Trinity College, Humphrey Lloyd*, to try to verify his theoretical prediction experimentally. This Lloyd did two months later, and this theoretical prediction brought great fame to Hamilton.

However, it unfortunately also led to controversy with his Dublin colleague, James MacCullagh*. MacCullagh had published some results on refraction three years earlier in papers which Hamilton had reviewed. MacCullagh published a note in which he claimed at least partial priority for the discovery of conical refraction. He wrote: "The indeterminate cases of my own theorems, which, optically interpreted, mean conical refraction, of course occurred to me at the time." However, although MacCullagh had come very close to the theoretical discovery himself, he was forced to admit that he had failed to take the last step.

On 4 November 1833 Hamilton read a paper to the Royal Irish Academy expressing complex numbers as algebraic couples, or ordered pairs of real numbers; this was a considerable step towards the modern approach. Instead of writing a complex number z as $x + \mathrm{i}y$, with i being the square root of minus one, and x and y called the 'real' and 'imaginary' parts of z, modern mathematicians prefer to write $z = \{x, y\}$, and to present, as axioms, rules for working with these ordered pairs.

In 1834, Hamilton presented a paper titled *On a General Method in Dynamics*. In this paper Hamilton applied his characteristic function, which until then he had used only in optics, to dynamics. Essentially his work was to provide a complete new mathematical basis to dynamics, an alternative to that of Newton. The approach, known as *Hamiltonian Mechanics*, is of great theoretical interest and importance, and can also be of considerable practical use in some problems.

Conceptually the method demonstrated a strong mathematical analogy between optics and mechanics. In optics, the simplest approach is via *geometrical optics*, which treats light as a ray travelling in a straight line in a single medium, but which is bent when it moves from one medium to another. Geometric optics may be obtained as an approximation from *physical optics*, which considers light more rigorously as a wave motion, and so can handle phenomena such as interference and diffraction, which cannot be understood using geometrical optics.

Similarly Hamiltonian mechanics relates the movement of a particle along a continuous line, which would be straight in the absence of any applied forces, to an underlying wavelike function, called the Hamiltonian. This fusion of particle-like and wavelike concepts meant that modern quantum mechanics, the starting-point of which is the problem of combining in some way particle-like and wavelike behaviour, could be constructed on the basis of Hamiltonian mechanics. Indeed Hamiltonian methods were at the centre of the celebrated development of quantum theory by Schrödinger*. Hamiltonian mechanics plays a fundamental part in today's quantum mechanics, a theory invented around 60 years after Hamilton's death.

Hamilton presented his arguments very tersely, and so his early papers are hard to read. His approach was different from that now commonly presented in textbooks describing the method. He defined the characteristic function V as the action of the system in moving from its initial to its final point in configuration space, and he made the initial and final coordinates the independent variables of V. For conservative systems, the total energy H was constant along any path, but it varied if the initial and final points were varied, and, for the n-particle case, V became a function of the $6n$ coordinates of initial and final position and of what is now called the Hamiltonian H.

In 1834, Hamilton and Helen had a son, William Edwin, but Helen then left Dunsink for nine months, leaving Hamilton to fight the loneliness by throwing himself into his work. In 1835 he published *Algebra as the Science of Pure Time*, which was inspired by his study of Kant, and he presented it to a meeting of the British Association for the Advancement

Irish half penny stamp celebrating the centenary of Hamilton's discovery of quaternions.

of Science. This second paper on algebraic couples identified them with steps in time and he referred to the couples as 'time steps'.

Hamilton was knighted in 1835 and that year his second son, Archibald Henry, was born, but the next few years did not bring him much happiness. After the discovery of algebraic couples, he tried to extend the theory to triplets, and this became an obsession that plagued him for many years. His family life also plummeted. The following autumn he went to Bristol for a meeting of the British Association, and Helen took the children with her to Bayly Farm for ten months. His cousin Arthur died, and not long after Helen returned from her mother's, she went away again to England, this time leaving the children behind. This followed the birth of a daughter, Helen Eliza Amelia. At this point, William became depressed and started to have problems with alcohol, so his sister came back to live at Dunsink.

Helen returned in 1842 when Hamilton was so preoccupied with the triplets that even his children were aware of it. Every morning they would inquire: 'Well, Papa can you multiply triplets?', but he had to admit that he could still only add and subtract them. Then, on Monday 16 October 1843, Hamilton was walking along the Royal Canal with his wife, to preside at a Council meeting of the Royal Irish Academy. Although his wife talked to him now and again, Hamilton hardly heard, for the discovery of the *quaternions*, the first non-commutative algebra to be studied, was taking shape in his mind: "And here there dawned on me the notion that we must admit, in some sense, a fourth dimension of space for the purpose of calculating with triples.... An electric circuit seemed to close, and a spark flashed forth."

Hamilton could not resist the impulse to carve the formulae for the quaternions

$$i^2 = j^2 = k^2 = ijk = -1$$

in the stone of Brougham Bridge as he and his wife passed it. Hamilton felt this discovery would revolutionize mathematical physics and he spent the rest of his life working on quaternions. He wrote: "I still must assert that this discovery appears to me to be as important for the middle of the nineteenth century as the discovery of fluxions [the calculus] was for the close of the seventeenth."

Shortly after Hamilton's discovery of the quaternions, his personal life started to prey on his mind again. In 1845, Thomas Disney visited Hamilton at the observatory and brought Catherine with him. This must have upset William, as his alcohol dependency took a turn for the worse. At a meeting of the Geological Society the following February he made an exhibition of himself through his intoxication. Macfarlane (in Further Reading) writes: "At a dinner of a scientific society in Dublin he lost control of himself, and was so mortified that, on the advice of friends he resolved to abstain totally. This resolution he kept for two years, when... he was taunted for sticking to water, particularly by Airy.... He broke his good resolution, and from that time forward the craving for alcoholic stimulants clung to him."

The year 1847 brought the deaths of his uncles, James and Willey, and the suicide of his colleague at Trinity College, James MacCullagh, which greatly disturbed him despite the fact that they had had severe disagreements.

The following year Catherine began writing to Hamilton, which cannot have helped at this time of depression. The correspondence continued for six weeks, and it became more informal and personal until Catherine felt so guilty that she confessed to her husband. Hamilton wrote to Barlow and informed him that they would never hear from him again. However, Catherine wrote once more, and this time her remorse was so great that she attempted suicide, fortunately unsuccessfully. She was to spend the rest of her life living with her mother or siblings, although there was no official separation from Barlow. Hamilton persisted in his correspondence to Catherine, which he sent through her relatives.

It is no surprise that Hamilton gave in to alcohol immediately after this, but he threw himself into his work, and began writing his *Lectures on Quaternions*. He published these lectures in 1853, but he soon realized that it was not a good book from which to learn the theory of quaternions. Perhaps Hamilton's lack of skill as a teacher showed up in this work.

At this time, Hamilton helped Catherine's son James to prepare for his Fellowship examinations which were on quaternions. He saw this as revenge towards Barlow, as he was able to help the boy in a way that his father could not. Later that year Hamilton received a pencil case from Catherine with an inscription that read: 'From one who you must never forget, nor think unkindly of, and who would have died more contented if we had once more met'. Hamilton went straight to Catherine and gave her a copy of *Lectures on Quaternions*. She died two weeks later.

As a way of dealing with his grief, Hamilton plagued the Disney family with incessant correspondence, sometimes writing two letters a day. Lady Campbell was another sufferer of the burden of mail, as only she and the Disneys knew of his love for Catherine. On the other hand, Helen must have always suspected that she did not take first place in her husband's heart, a notion that must have been strengthened in 1855 when she found a letter from Dora Disney, Catherine's sister-in-law. This led to an argument, although the only consequence was that Dora had her letters addressed by her husband; they did not stop altogether.

Determined to produce a work of lasting quality, Hamilton began to write another book, *Elements of Quaternions*, which he estimated would be 400 pages long and take two years to write. The title makes it clear that suggests that Hamilton modelled his work on Euclid's *Elements*. The book took seven years to write, and ended up double its intended length. In fact the final chapter was incomplete when he died, and the book was finally published with a preface by his son William Edwin Hamilton.

Not everyone found Hamilton's quaternions the answer to everything they had been looking for. Lord Kelvin* wrote: "Quaternions came from Hamilton after his really good work had been done, and though beautifully ingenious, have been an unmixed evil to those who have touched them in any way". Cayley compared the quaternions with a pocket map "which contained everything but had to be unfolded into another form before it could be understood".

Hamilton died from a severe attack of gout, shortly after receiving the news that he had been elected the first foreign member of the National Academy of Sciences of the USA.

Further Reading

Article by Thomas Hankins in: *Dictionary of Scientific Biography* (C G Gillespie, ed) (Scribner, New York, 1970–80).

Life of Sir William Rowan Hamilton, Robert P Graves (Hodges, Figgis and Co., Dublin, 1882, 3 volumes).

Sir William Rowan Hamilton, Thomas L Hankins (Johns Hopkins, Baltimore, 1980).

William Rowan Hamilton: Portrait of a Prodigy, Seán O'Donnell (Boole, Dublin, 1983).

Lectures on Ten British Mathematicians of the Nineteenth Century, Alexander Macfarlane (Wiley, New York, 1916) pp 34–49.

'The Significance and Development of Hamilton's Quaternions', H T H Piaggio, *Nature* **152** (1943) 553–555.

James MacCullagh

1809–1847

James G O'Hara

The mathematical physicist James MacCullagh was born in 1809—the exact day and month are unknown—in Landahussy, a townland in the parish of Upper Badoney near Plumbridge in County Tyrone. He was the eldest of 12 children—four of whom died young—of James MacCullagh (1777–1857) and his wife, Margaret MacCullagh (née Ballentine [?], 1784–1839). While he was still young his father left his mountain farm and the family moved to Strabane, where the mathematical talent of the young James first became apparent. He attended the schools of the historian Rev John Graham (1776–1844) and the classicist Rev Thomas Rollestone (1789–1850) at Lifford in County Donegal.

On 1 November 1824 MacCullagh was admitted to Trinity College Dublin, and on 1 June 1825 he was a successful candidate for a scholarship or sizarship, as it was known. He took up residence in the College in June 1826, and he lived there until the day of his death. He was elected a Foundation Scholar of the College in June 1827, and graduated as a Bachelor of Arts in March 1829. MacCullagh benefited greatly from the renaissance in mathematical education at the College initiated there in the 1820s by Bartholomew Lloyd, and his rapid promotion owed much to Lloyd's patronage.

He strove for an academic career from the outset, but in his first two attempts at the highly competitive Fellowship examination in June 1829, and again in 1831, he was unsuccessful. He then sent Lloyd, who had become Provost, a letter containing several theorems relating to a geometrical theory of rotation, and to the theory of attraction. The publication in 1834 of Louis Poinsot's theory of rotary motion meant that MacCullagh's work had been superseded, but his letter was published in 1844 in the *Proceedings of the Royal Irish Academy* by the Provost's son, the physicist Humphrey Lloyd*.

MacCullagh competed successfully for a Fellowship on his third attempt in 1832; this opened the door for him to an academic career at the College and he was elected a Junior Fellow on 18 June 1832. Shortly afterwards he was made Junior Assistant to the Erasmus Smith Professor of Mathematics, and Assistant to the Lecturer in Greek. Until 1840 Fellows

Sketches of James MacCullagh made after his death by F W Burton (courtesy of the National Gallery of Ireland).

were required to be celibate, but, in any case, there is no evidence that he ever contemplated matrimony; he would remain a lifelong bachelor.

In the course of the 18 years of his career, MacCullagh wrote some 39 papers, most of which were presented to the Royal Irish Academy, and were published in the Academy's *Proceedings* or *Transactions*. Most of his papers were republished in 1880, together with his lectures on the rotation of a solid body, and on the attraction of ellipsoids, in *The Collected Works of James MacCullagh*. Of his first two papers, which were published in 1830, one was devoted to geometrical theorems on the rectification, that is to say, the determination of the arc-length, of conic sections, and the other dealt with double refraction in a crystallized medium.

In this second paper he provided a description of the so-called 'Fresnel wave surface'. Long before Fresnel's day—Augustin Jean Fresnel lived from 1788 to 1827—the Dutchman Christiaan Huygens (1629–1695) showed that the explanation of optical phenomena might be based on the concept of such a wave surface, not only in isotropic media, like glass or water, in which it has a spherical form, but also in certain crystals for one of which, Iceland spar, he deduced the form of the surface. Light passing through such crystals was observed to be doubly refracted, that is, an extraordinary ray is transmitted alongside the normal or ordinary refracted ray.

The ray-surface or wave surface, which represents the distance traversed by the rays during a given interval of time in various directions from a point of origin within the crystal, consists, in crystals like Iceland spar and quartz, of two sheets, namely a sphere (corresponding to the ordinary ray) and an ellipsoid of revolution (corresponding to the extra-ordinary ray). The sphere and the ellipsoid are concentric and touch at the ends of their common axis, which corresponds to a single direction for which no double refraction is observed.

In the early nineteenth century Sir David Brewster discovered a further class of crystals, having two axes of no double refraction, and they were called biaxial crystals. An example is aragonite, a mineral form of calcium carbonate ($CaCO_3$). Here the wave surface consists of two sheets that interpenetrate giving a complex mathematical figure. This surface was first investigated by Fresnel, but his description remained incomplete.

MacCullagh's colleague William Rowan Hamilton* also studied Fresnel's wave surface for biaxial crystals, and predicted the existence of a new type of 'conical refraction' for light entering and leaving such crystals. Hamilton saw that there were four trumpet-like cusps or indentations at the points where the sheets of the wave surface interpenetrate. Rays corresponding to the lines drawn from the origin to these conical points would be refracted as cones of rays, a totally unexpected phenomenon that was demonstrated experimentally by Humphrey Lloyd at Trinity College Dublin in December 1832.

The discovery of conical refraction was regarded as a decisive proof of the wave theory of light, and a blow to the rival particle theory, which had no explanation for the phenomenon. In announcing the discovery to the Academy, Hamilton failed, however, to refer to MacCullagh's paper on the Fresnel wave surface, and this prompted the latter to publish a note in the *Philosophical Magazine* in August 1833, in which he showed how the phenomenon could be deduced from his theorems. Hamilton was greatly peeved but, following the intervention of Humphrey Lloyd, MacCullagh was persuaded to publish a further note in September, explaining that his former note had been written in haste, and that his deduction of conical refraction was subsequent to Hamilton's discovery. Evidence of MacCullagh's profound understanding of double refraction in crystals is evident in his first major paper *Geometrical Propositions Applied to the Wave Theory of Light* published later in 1833.

MacCullagh's most important work in physics was in the field of theoretical crystallography; he set about trying to improve the derivation by Fresnel of the laws governing the reflection of light, and to extend them to crystalline surfaces. For a hypothetical ether in the form of an elastic solid having the same density everywhere, he assumed, unlike Fresnel, that the vibrations were parallel to the plane of polarization, and, like Fresnel, that the incident transverse waves give rise to transverse reflected and refracted waves. MacCullagh's theory of crystalline reflection and refraction (1835–1837) was identical with a theory of the German Franz Ernst Neumann announced in 1835 and published in 1837. In papers of

1836 MacCullagh treated the optical behaviour of quartz and the reflection of light from metals.

MacCullagh's most important paper on light, *An Essay towards a Dynamical Theory of Crystalline Reflexion and Refraction*, in which he set forth equations which describe a light-bearing ether having the properties necessary to justify the assumptions he had made in earlier work on crystalline reflection, was read on 9 December 1839, but the final version was not published until 1843. In this work he followed the method adopted by the English mathematical physicist, George Green, in an 1838 investigation of the propagation of waves in a real elastic medium. Early in 1840, MacCullagh was able to reconcile his theory of optical activity in quartz with his dynamical theory, and, in May 1841, he gave an account of how his theory could be extended to include total reflection. MacCullagh was not an experimental physicist, but in 1843 he published the results of experiments carried out in 1837 on metallic reflection with the assistance of the Dublin instrument maker Thomas Grubb*.

The 1820s and 1830s had provided—through the experimental investigations of Thomas Young, Fresnel and Humphrey Lloyd on interference phenomena—major new evidence in support of the wave theory. This theory now gained the upper hand over the emission or particle theory that had been dominant since the time of Newton. MacCullagh remained sceptical, however, about the ultimate truth of the wave theory. At the annual meeting of the British Association for the Advancement of Science at Manchester in 1842, where a vigorous debate between protagonists and antagonists of the wave theory took place, he adopted an agnostic stance and suggested the theory still lacked a sufficient foundation on physical principles.

Although MacCullagh's own work had provided a mathematical framework for the description of a wide range of optical phenomena, his work was received with scepticism by many contemporaries. His dynamical theory did, however, find supporters, particularly among the Anglo-Irish, decades after his death. In 1880 George Francis FitzGerald* provided an interpretation of his theory of reflection and refraction so as to bring it into harmony with James Clerk Maxwell's electromagnetic theory. Though George Gabriel Stokes*—who hailed from the west of Ireland and became one of the leading physicists in Britain—was particularly critical of MacCullagh's work, Joseph Larmor* and Lord Rayleigh considered this judgement too harsh. Larmor spoke of MacCullagh as one of the great figures of optics.

MacCullagh's primary talent was as a mathematician and his success in developing physical theory was due in part to his geometrical skills. He had been interested in surfaces, especially the ellipsoid and second-order surfaces, since 1829, when he first investigated the Fresnel wave surface. The ellipsoid was also relevant to his early investigations of the rotation of a solid body. A form of geometry later known as inversive geometry, and the use of reciprocal surfaces, were used to good advantage, both in his elucidation of the Fresnel wave surface, and in his treatment of the rotation of a solid body. This idea was then extended in a major mathematical work of 1843, *On Surfaces of the Second Order*, to what is referred to as the modular generation of surfaces such as the ellipsoid. As with all his mathematical writing, this paper is characterized by its elegance and simplicity of style.

On 23 November 1835 the Chair of Mathematics at Trinity College Dublin became vacant, and shortly afterwards MacCullagh was appointed. In the years that followed, he received a variety of honours. In the summer of 1838 he was conferred with the legal degrees LLB and LLD. He had previously been elected a member of the Royal Irish Academy in February 1833. Now he was elected to the Academy's council and was awarded its Cunningham medal, for his paper on the laws of crystalline reflection and refraction, the presentation being made by Hamilton on 25 June 1838. On this occasion Hamilton referred to Neumann's paper, but gave priority to MacCullagh. The pronouncement had repercussions as far away as Germany. MacCullagh's claim was disputed by Neumann in a letter of 8 October, which was read at the meeting of the Royal Irish Academy on 30 November. MacCullagh insisted on his priority in publication and the independence of his work.

MacCullagh made an important contribution to the development of the School of Mathematics at Trinity College Dublin and helped establish a geometrical bias there. He delivered a special course of lectures to the fellowship candidates, and from 1837 to 1843 was an examiner at the annual fellowship examinations. He was an inspiring lecturer and of his graduate students after 1835, twenty became fellows and a number were to make original contributions in mathematics or physics. He was instrumental in the establishment of the School of Engineering in 1841, and he subsequently shared responsibility for teaching mechanics and physics to engineering students. On 4 December 1843 he was appointed Erasmus Smith Professor of Natural and Experimental Philosophy, or Physics, in succession to Humphrey Lloyd, and he was allowed to substitute a course of physics lectures in place of those he had previously provided in mathematics for fellowship candidates.

MacCullagh's other scholarly interests included Egyptian chronology, on which he read a paper to the Royal Irish Academy in 1837. Following a correspondence in 1842 with the orientalist Edward Hincks, he published a second paper, and was instrumental in obtaining the services of Hincks to undertake a descriptive catalogue of the College collection of Egyptian papyri which was published in 1843. Irish culture likewise figured among his interests. He devoted much time and effort to building up the Royal Irish Academy's Museum of Antiquities, which is now part of the National Museum of Ireland. He purchased at his own expense the early twelfth-century Cross of Cong for the Academy, raised a fund to purchase two gold torcs (c. 1200 BC) found at Tara, and provided £300 for the purchase of the Domnach Airgid (Silver Shrine).

On 22 October 1842 the Committee of Physics of the Royal Society of London, under the chairmanship of Humphrey Lloyd, recommended that the Copley medal for that year be awarded to MacCullagh. The presentation, on 30 November 1842, was made on his behalf to his friend the physicist Charles Wheatstone. On 2 February 1843 his election as a Fellow of the Royal Society followed. MacCullagh attended the annual meetings of the British Association for the Advancement of Science in Dublin (1835), Bristol (1836), Manchester (1842) and Cork (1843). He travelled to the continent on at least three occasions (in 1840, 1842 and 1846). Together with Charles Babbage, the English pioneer of computer science, he attended the meeting of the Italian Physical Society at Turin in September 1840. With the support of Babbage and Wheatstone he was elected to the London Athenaeum Club in February 1842.

The last months of MacCullagh's life were marked by an involvement in politics. In the general election of 1847 he decided to compete for one of the two Dublin University seats at Westminster, opposing the sitting members who were Oxford graduates. MacCullagh was a liberal, but without party affiliation or political experience, and in a constituency that was strongly Tory he was an outsider from the outset. In an election address of 25 June 1847 he pledged to promote the interests of the University and the Established Church, and to work for the development of the country's natural resources, and the improvement of the condition of the people. The latter pledge suggests perhaps a reaction from the trauma of the Great Famine. As he was one of three lay Fellows of the College—he had been freed on 20 February 1836 at his own request from the requirement to take holy orders in the Church of Ireland—there may have been doubts about his religious orthodoxy.

In the election that took place from 4 to 9 August 1847 he received 374 out of 2224 votes cast, finishing last of four candidates. The Young Irelanders applauded his patriotism after his death, an obituary notice in *The Nation* (Saturday 30 October 1847) describing him as a warm and ardent nationalist. It is likely, however, that he was a nationalist only in the sense that he believed Irishmen should show self-respect and promote their own national institutions. He made no pronouncements on constitutional issues such as the Union with Great Britain.

MacCullagh's powerful drive for achievement was frustrated by a series of disappointments in science and politics. His psyche was characterized by excessive sensitivity, introspection and a lack of stability. An impetuous temperament combined with suspicion or fear of plagiarism caused him to make extravagant claims on occasions. He died on the

Bust of James MacCullagh by Christopher Moore (courtesy of the provost, fellows and scholars of Trinity College Dublin).

evening of Sunday 24 October 1847, being found in his College apartment with his throat cut. An inquest the next day returned a verdict of suicide. The verdict has never been seriously questioned, but a certain mystery did arise from the fact that no trace of his collection of manuscript papers was found after his death.

A funeral service was held in Trinity College on 30 October, and the remains were buried in the family vault in the graveyard of St Patrick's parish church of Upper Badoney, County Tyrone. He is commemorated by a brass plaque in that church, and by a marble tablet on the family grave erected by his sister Isabella (1823–1894). Three sisters and a younger brother were dependent on him at the time of his death, and an appeal to the Prime Minister, with the support of important academics and politicians, helped procure a Civil List pension for his sisters. In Trinity College Dublin, MacCullagh is honoured by a marble bust in the Common Room, carved in 1849 by Christopher Moore (1790–1863). In the National Gallery of Ireland there is a series of sketches by Frederick William Burton (1816–1900) drawn from memory immediately after MacCullagh's death. In conclusion, one of the most sincere tributes to him, that paid by his colleague, the mathematician and poet William Rowan Hamilton, who composed the following lines on the occasion of his death, may be aptly cited:

> 'Wrapped as we are in an overwhelming cloud
> Of grief and horror, shake we off awile

That horror, and that grief with words beguile;
And from our full hearts breathe, though not aloud.
Our minds to God's mysterious dealings bowed,
And mourning with the Genius of the land,
Take we awhile our reverential stand,
In the dread presence of MacCullagh's shroud
[*In the manuscript*: Beside MacCullagh's blood-bespotted shroud]
Great, good, unhappy! For his country's fame,
Too hard he toiled; from too unresting brain
His arachnæan web of thought he wove.
The planet-form [i.e. the ellipsoid] he loved, the crystal's frame
Through which he taught to trace light's tremulous train'.

Further Reading

Dictionary of National Biography and the forthcoming *New Dictionary of National Biography*.

'James MacCullagh, M.R.I.A., F.R.S., 1809–47', B K P Scaife, *Proceedings of the Royal Irish Academy* **90C** (1990) 67–106.

'James MacCullagh', T D Spearman in: *Science in Ireland 1800–1930: Tradition and Reform*, J R Nudds *et al* (eds) (Trinity College Dublin Press, 1988), pp 41–59.

'The Prediction and Discovery of Conical Refraction by William Rowan Hamilton and Humphrey Lloyd (1832–1833)', J G O'Hara, *Proceedings of the Royal Irish Academy* **82A** (1982) 231–257.

Sir William Rowan Hamilton, Thomas L Hankins (John Hopkins, Baltimore, 1980).

Thomas Andrews

1813–1885

D Thorburn Burns

Thomas Andrews was born on 19 December 1813, the eldest son of Thomas John Andrews, a linen merchant, and Elizabeth Stevenson at 3 Donegall Square South, Belfast. He lived there as a boy and man until his transfer to his College residence in 1845. His first school was Belfast Academy in Donegall Street but he was soon transferred to the Academical Institution.

After a short period working in his father's office during 1828 he then, at 15, went to study chemistry at Glasgow under Thomas Thomson. Two years later, in 1830, he went to Paris studying under André Dumas. Illness compelled his return to Ireland; he then trained at Trinity College Dublin as a medical student for four years. Next, he spent a year in Edinburgh, graduating MD, and also qualifying at the Royal College of Surgeons in 1835, before returning to Belfast.

Andrews set up in medical practice as a physician and was soon appointed Professor of Chemistry in the new Medical College at the Academical Institution. He combined medical practice with research and teaching chemistry. In 1842 he married Jean Hardie, daughter of Major Walker of the 42nd Highlanders. He gave up these occupations when he was appointed Vice-Principal of the projected Northern College of the Queen's University of Ireland in 1845. The Chair of Chemistry was officially founded in 1849 but Andrews was named first Professor in 1847 and remained in this post until retirement in 1879. His various dealings with the College are recorded in Moody and Beckett's history of the College (listed in Further Reading). Following his early educational visit Andrews travelled extensively in Europe, and met, and maintained contact with, many of Europe's leading chemists (see Table 1).

Upon retirement he moved to Fort William Park, Belfast. After a period of ill health he died on 26 November 1885, and was buried at the Borough Cemetery, Belfast, where a granite obelisk marks the grave.

Andrews' scientific papers may be found in the collection of Tait and Crum Brown, and in the Royal Society catalogues of papers, both listed in Further Reading. His work may be divided into five main groups, namely: (i) electrochemical studies, (ii) thermochemistry, (iii)

Table 1. Andrews' travels and contacts with Europe's leading chemists.

Years	Locations	Scientific contacts
1830–31	Paris	Thenard, Dumas
1836	Paris	Dumas, Graham, Cheveral, Berthier, Gay-Lussac
1848	Paris	
1850	Basle, Vienna, Verona, Venice, Prague, Dresden, Hanover, Cologne, Brussels	Schonbein, Schrotter
1854	Heidelberg, Munich	Bunsen, Delffs, Rau, Liebig, Hoffmann
1856	Paris	Dumas, Despretz, Peligot, De Luca, Cheveral, Faraday, Grove
1875	Paris, Nantes, Lyons, Geneva	

critical states and liquefaction of gases, (iv) the nature of ozone, and (v) development of a series of analytical chemical methods.

Between 1841 and 1848 he paid particular attention to, and became extremely skilled at, thermochemical measurements. Many of his results for the heats of chemical reactions were remarkably accurate for the period, for example −13.1 kcal per mole for the reaction between sodium hydroxide and hydrochloric acid compared with the present-day accepted value of −13.4 kcal per mole (at infinite dilution). Andrews' work on this topic is discussed by Mackle (Further Reading).

Thomas Andrews in 1875.

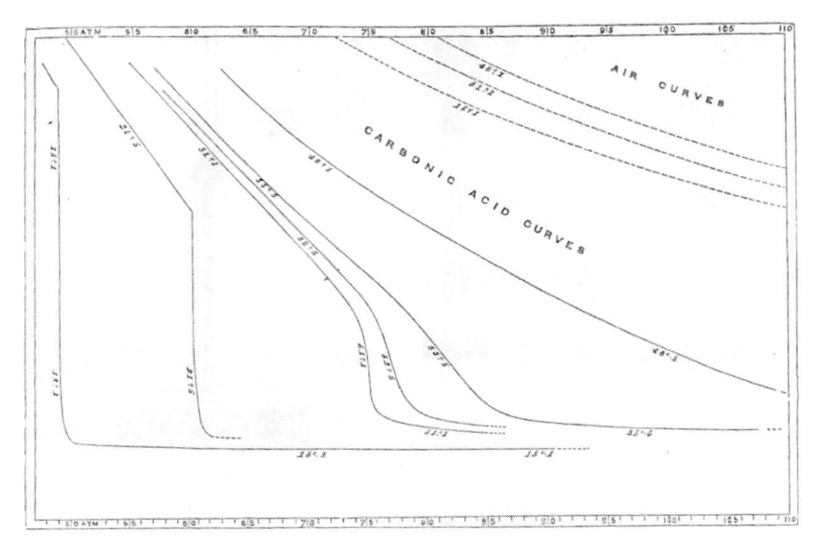

Andrews *P-V-T* diagram for carbon dioxide (Phil. Trans. (1869), 575).

He then turned his attention to the contemporary problem of the constitution of ozone. After a series of experiments he concluded that the supposed different sorts of ozone were identical. Attempts by Andrews and Tait to determine the density of ozone failed. This was because they did not understand that the reagents that removed ozonic properties from a mixture of ozone and oxygen, combined with ozone, removed an atom of oxygen from a molecule of ozone leaving a molecule of oxygen. Hence no change in volume took place. Since a measurable quantity of ozone thus appeared to occupy zero volume, it appeared that its density was infinite. This perplexing result led to the true solution by Odling in 1861 and the formula of O_3 for ozone. (See the paper by Thorburn Burns listed in Further Reading.)

Andrews is best known for his now classical work, that on the continuity of the gaseous and liquid states, particularly for his discovery of the critical temperature of carbon dioxide in 1861. Andrews' value was 30.92°C, the modern accepted value being 31.04°C. These studies formed the basis of his Bakerian Lectures in 1869 and 1876. The *P-V-T* diagrams constructed by subjecting carbon dioxide and a 'permanent gas' to the same pressure (calculated from the latter's volume) and temperature are known to generations of students as 'Andrews isotherms'.

In the experiment, carbon dioxide and air were sealed in separate capillaries and subjected to pressure by mercury compressed by a screw. A tube connected the two sides. Scales were attached and the capillaries and scales surrounded by water baths. Then, as now, money for research was scarce; Andrews recorded his *P-V-T* data on the back

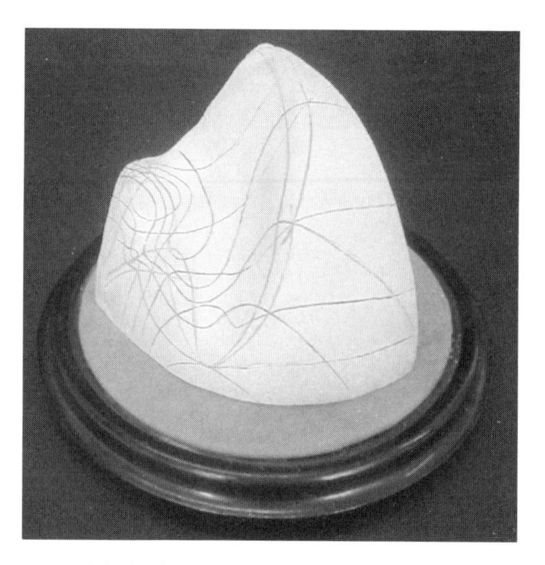

Clerk Maxwell's own model of a thermodynamic surface showing the relationship between energy, entropy, and volume for CO_2. This model was sent to Andrews by Clerk Maxwell in 1875 after Andrews had given his first Bakerian Lecture in 1869 (courtesy of Queen's University, Belfast).

of students' examination papers that had been used only on one side. The subsequent rigorous thermodynamic analyses and interpretations in terms of intermolecular forces was discussed in 1969, the centenary of the 1869 Bakerian Lecture, by Rowlandson (see Further Reading).

Andrews worked mainly alone, but had the assistance of P G Tait for part of the work on ozone. He was noted for his manipulative skills and his ingenuity in solving difficult practical problems, and he constructed most of his own apparatus such as thermometers and the dividing engine for their calibration, the complex glassware for the experiments with ozone, and so on.

In his obituary listed in Further Reading, Crum Brown noted: "[While] this independence and individuality limited the amount of work done by him, it has the compensating advantage that we know that every analysis and every observation published by him were actually made with his own hands and eyes, so that a reader of his paper is as nearly as possible in as good a position to judge the soundness of his conclusions as if he had performed the experiments himself. In reading his papers we are transported at once to the laboratory; without wearisome repetition we have all the details before us, and we can follow every step of his argument as if we had been present at every experiment on which it was founded."

From student days onwards Andrews received many distinctions, honours and awards. These are summarized in Table 2. Noteworthy

Table 2. Andrews' distinctions, honours and awards.

Year	Distinction/award	Society/university
1839	Member	Royal Irish Academy
1841	Foundation Member	Chemical Society, London
1844	Royal Medal	Royal Society, London
1846	Honorary Member	Société des Sciences Naturelles du Canton de Vaud
1849	Fellow	Royal Society, London
1850	1000 Franc Prize	Academy of Science (Paris)
1852	President of Chemical Section	British Association, Belfast Meeting
1869	Bakerian Lecture	Royal Society, London
1870	Honorary Fellow	Royal Society of Edinburgh
1871	LLD	Edinburgh University
1871	President of Chemical Section	British Association, Edinburgh Meeting
1873	LLD	Trinity College Dublin
1875	Vice-President, Chemical Section	French Association for Advancement of Science, meeting at Nantes
1876	Bakerian Lecture	Royal Society, London
1876	President	British Association, Glasgow Meeting
1877	LLD	Glasgow University
1879	DSc	Queen's University, Ireland
1880	Declined offer of knighthood	Queen Victoria
1884	Corresponding Member	Royal Society of Sciences of Göttingen

was the award of the Royal Medal, in the gift of the Royal Society of London, in 1844, for his work on 'thermal changes accompanying basic substitutions'. In 1849, the same year as his election to the Royal Society, he was given a prize by the Academy of Sciences, Paris, for his 'memoir on the determination by the quantity of heat disengaged in chemical combination'. This was given a *titre d'indemnité*, as the greatest part of its contents had already been published since its submission in 1845 due to the slowness of the adjudication process.

In 1880 Andrews received a letter from the Duke of Marlborough, then Lord Lieutenant of Ireland, offering him, by Her Majesty's gracious permission, the honour of civil knighthood. In the memoir by Tait and Crum Brown, it is stated that 'Dr Andrews, though appreciating deeply the kindness of purpose, declined the distinction on grounds of indifferent health'. This excuse has been given in most accounts of his life published since. The position may, though, not be so simple, since in the obituary in *Nature* we read: "He received honorary degrees from various Universities. But he valued this class of distinctions simply as tokens of the esteem and good wishes of the donors; and in the somewhat delicate matter of a civil title he shared the opinion, and followed the practice of his cherished friend Faraday'' (Faraday's religious views did not allow him to accept worldly honours.) Regrettably, the name of the author of this obituary was not recorded.

Andrews was deeply interested in political and social affairs but rarely took an active part in politics and was free from party feelings. His writings in this area were called 'Chapters in Contemporary History'. The first, 'Studium Generale' written in 1867, is a detailed historical and critical discussion of the function of a university, with special reference to the Queen's Colleges. The second more contentious volume, 'The Church in Ireland' written in 1869, started with a historical sketch of religious bodies in Ireland, and was a plea in favour of disestablishment of the Church of Ireland and the equitable distribution for spiritual purposes of the church property amongst the whole population.

As President of the Social Science Congress which met in Belfast in 1867, Andrews read a paper, entitled *Suggestions for Checking the Hurtful Use of Alcoholic Beverages by the Working Classes*. He suggested that no house should be licensed as a public house for the sale of alcoholic beverages unless it be provided with ample appliances for cooking and serving food; and that no licensed publican be allowed to sell, or keep in store, any liquor containing 17% of alcohol, a figure to be reduced in due course to 12%, the burgundy standard.

Andrews took an active part in the dissemination and popularization of science for the general public. He joined the Belfast Natural History and Philosophical Society in 1835, and over the years contributed 27 papers to its meetings on a range of topics relating to his research and, more generally, based on the works of others. Riddel has written an account of Andrews' interaction with Society; it is cited in Further Reading. He was also a keen member of the British Association for the Advancement of Science. Andrews contributed papers to many meetings of the Association, as well as being President of the Chemical Section at Belfast in 1852, and again in Edinburgh in 1871, and overall President at Glasgow in 1876.

Although 'Studium Generale' was in many ways forward looking, this was not true of all his views. For example, he is recorded as making a telling speech in 1873, at the Convocation of the Queen's University in Ireland, against extending the privileges of the university to women. Earlier, in 1870, at the Council of Queen's College, Belfast he proposed ''that women should be allowed to attend lectures if the professors concerned considered it expedient, and were satisfied that the discipline of the classes would not suffer''.

All accounts of Andrews' life note his characteristic of hard work, that he was warmly hospitable, indeed personally almost stoically temperate, allowing himself the minimum of rest, of food and of sleep. In

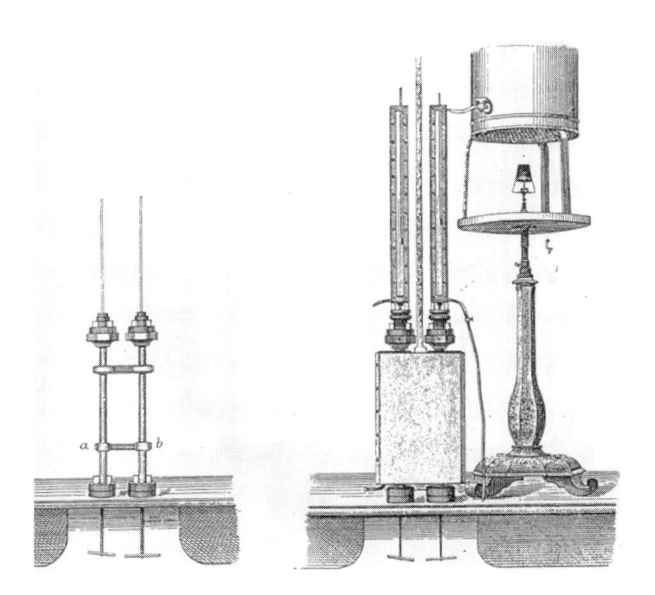

Apparatus for studying the continuity of the liquid and gaseous states (Phil. Trans. (1869), 575).

theological matters, as in all others, he thought for himself; he was an excellent example of the true Christian philosopher.

Andrews via his work on the continuity of the liquid and gaseous states and the liquefaction of gases paved the way for the development of many aids to current everyday life such as refrigerators, as well as to the use of low temperatures, essential to much modern work in physics and chemistry.

Upon his retirement, the Queen's College, Belfast, proposed to commemorate his services by a portrait to be placed in the College and a replica for the family, and a prize or scholarship. The Andrews' Studentship for the promotion of Chemical and Physical Science was established in 1993 and continues to this day; the official portrait hangs in the University Great Hall.

Further Reading

Obituaries of Andrews appear in *Nature* **33** (1885) 157; *Journal of the Chemical Society* **49** (1886) 342; *Proceedings of the Royal Society of London* **41** (1886) xi.

The Scientific Papers of the Late Thomas Andrews MD FRS with a Memoir, P G Tait and A Crum Brown (Macmillan, London, 1889).

Article in *Dictionary of National Biography*.

'Dr. Thomas Andrews: The Great Chemist and Physicist', *Proceedings of the Belfast Natural History and Philosophical Society* **107** (1920–1921).

'Schools of Chemistry in Great Britain and Ireland XXIX—The Queen's University of Belfast', *Journal of the Royal Institute of Chemistry* **17**, (1957) 81.

'Thomas Andrews: Physical Chemist, Physician and University Administrator', William Davis, in: *Some People and Places in Irish Science and Technology*, Charles Mollan, William Davis and Brendan Finucane (eds) (Royal Irish Academy, Dublin, 1985), pp 46–47.

'Thomas Andrews (1813–1885)', H Mackle and C L Wilson, *Endeavour* **8** (1971) 30.

Queen's Belfast 1845–1949, T W Moody and J C Becket (Faber, London, 1959, 2 volumes).

Royal Society Catalogues of Scientific Papers (HMSO, London, 1867; Murray, London, 1879; Cambridge University Press, 1902 and 1915).

'Thomas Andrews, Calorimetrist', H Mackle, *Nature* **224** (1969) 543.

Historical Aspects of the Analytical Chemistry of Ozone, D Thorburn Burns in: *The Chemistry of the Atmosphere, Oxidants and Oxidation in the Earth's Atmosphere* (Royal Society of Chemistry, London, 1995).

'Thomas Andrews and the Critical Point', J S Rowlandson, *Nature* **224** (1969) 543.

Thomas Andrews' general writings consisted of *Studium Generale: A Chapter of Contemporary History*, and *The Church in Ireland: A Second Chapter in Contemporary History*, both published by Longmans, Green and Co. in London in 1867 and 1869 respectively; and *Suggestions for Checking the Hurtful Use of Alcoholic Beverages by the Working Classes*, Transactions of the National Association for the Promotion of Social Science, Belfast Meeting, 1867, p 89.

George Gabriel Stokes

1819–1903

Alastair Wood

The contributions of G G Stokes to mathematics and physics have been so many and varied that it is difficult to know where to begin. A contemporary of Kelvin* and Maxwell, his name has become well known to generations of scientists, mathematicians and engineers, through its association with various physical laws and mathematical formulae. In standard textbooks of mathematics, physics and engineering we find Stokes' drift, Stokes' Law, Stokes' Theorem, Stokes' Phenomenon, Stokes' conjecture and the Navier–Stokes equations.

But while Stokes' contributions to mathematical physics are impressive, many believe that his greater contribution was as a sounding board for his contemporaries, providing sound advice, good judgement and mathematical rigour. Many great scientists deliberately avoid administration, committees and editorship of journals on the grounds that such activities would stifle their creativity. But Stokes threw himself into this role and, although a shy man of few words, was regarded with affection by colleagues throughout Britain and Ireland for his extraordinary generosity in encouraging their work and communicating ideas to others, usually through his extensive correspondence.

George Gabriel Stokes has long been associated with the University of Cambridge, where he spent all of his working life, occupying the Lucasian Chair of Mathematics, once held by Isaac Newton, from 1849 until his death in 1903. In this he was like William Thomson, later Lord Kelvin, who is sometimes associated with Glasgow (where he occupied the chair of Natural Philosophy), rather than with Belfast, where he was born. Stokes was born in Skreen, County Sligo, where his father was Rector of the Church of Ireland, and he received his early education there from the Parish Clerk, and later at Dr Wall's School in Dublin. Although his birthplace, the old Rectory, is now demolished, a memorial was erected at the site in 1995, and there is a signpost from the main Sligo to Ballina road. But the Parish Church, which contains a plaque to the Rev Gabriel Stokes, his father, is still in use, and the former Parochial School is now the Church Hall and the location of the biennial Stokes Summer Schools which draw participants from all parts of the globe.

Stokes in 1857 at the age of 38. From J Larmor (ed) 1907 *Memoir and Scientific Correspondence of the Late Sir George Gabriel Stokes*, 2 vols (Cambridge: Cambridge University Press) vol. I frontispiece.

The Stokes family is recorded in the English county of Wiltshire from the twelfth century, moving to Ireland early in the seventeenth century, although the exact date is unknown. The first of the Stokes family to be mentioned in Ireland was Gabriel Stokes, born in 1682, a mathematical instrument maker residing in Essex Street, Dublin, who became Deputy Surveyor General of Ireland. He is credited with having designed Pigeon House Quay in Dublin. Some readers will know this as the long stone jetty which runs out to the modern Poolbeg power station on the south side of Dublin Port. Among his other concerns was the use of 'hydrostatic balance' to ensure a piped water supply to Dublin. His great grandson, George Gabriel, returned to this problem in one of his earliest papers, *The Internal Friction of Fluids in Motion*, where he discussed an application to the design of an aqueduct to supply a given quantity of water to a given place. Gabriel's elder son, John, was Regius Professor of Greek, and his younger son, another Gabriel, was Professor of Mathematics, both in Dublin University. The descendants of this professor of mathematics became an important medical family in Ireland and internationally. Their name is preserved in medicine through Cheyne-Stokes respiration and the Stokes-Adams syndrome in cardiology. It is interesting to note that George Gabriel, while primarily a mathematical physicist, crossed the boundary between mathematics and medicine by discovering the respiratory function of haemoglobin.

It is from the first Gabriel's elder son, John Stokes, that George Gabriel Stokes is descended. Much less is known about his branch of the family. In 1798, the Rev Gabriel Stokes, son of John Stokes, and Rector of Skreen, County Sligo, married Elizabeth, the daughter of John Haughton, the Rector of Kilrea. Their first child, Sarah, died in infancy, but they produced seven further children, of whom George Gabriel was the youngest. All of his four brothers became clergymen, the oldest, John Whitley, who was already 20 when George Gabriel was born, becoming Archdeacon of Armagh.

In later life Stokes talked fondly of the scenery of his boyhood, and his rambles within sound of the Atlantic breakers. Even in his paper, *On the Theory of Oscillatory Waves*, he writes, in the midst of mathematical equations, of 'the surf which breaks upon the western coasts as the result of storms out in the Atlantic'. This very private and reserved Victorian scientist had the occasional habit of breaking into poetical descriptions in the middle of mathematical proofs. In his 1902 paper on asymptotics, he describes what is now known as Stokes' phenomenon as "the inferior term enters as it were into a mist, is hidden for a little from view, and comes out with its coefficient changed". Perhaps as a boy he had watched the mists skim the surface of flat-topped Benbulben across the bay, an area which was later to influence the poet W B Yeats, who is buried in nearby Drumcliffe.

There can be no doubt that George Gabriel was greatly inspired by his upbringing in the West of Ireland, and he returned regularly for the summer vacation while a student in England, a non-trivial exercise in the pre-railway era. Even after the death of his parents, he continued to visit his brother John Whitley, then a clergyman in Tyrone, and his sister, Elizabeth Mary, to whom he was greatly attached, at 4 Windsor Terrace, Malahide (where there is a plaque to him in St Andrew's Church) almost annually until his death.

The Rev Gabriel Stokes died in 1834, and his widow and two daughters had to leave Skreen Rectory, but money was found to send George Gabriel to continue his education at Bristol College in England. His mathematics teacher, Francis Newman, brother of Cardinal Newman, wrote that Stokes "did many of the propositions of Euclid as problems, without looking at the book". Stokes appears to have had a great affection for Newman, whom he records as having "a very pleasing countenance and kindly manners". Newman was also responsible for the following anecdote quoted in Larmor's 1907 Memoir: "His habit, often remarked in later life, of answering with a plain yes or no, when something more elaborate was expected, is supposed to date from his transference from an Irish to an English school, when his brothers chaffed him and warned him that if he gave long Irish answers he would be laughed at by his school fellows''.

George Gabriel Stokes entered Pembroke College, the third oldest in Cambridge, as an undergraduate in 1837. So effective were his studies that Stokes was Second Wrangler (that is, placed second in Part II of the Mathematical Tripos) in 1841 and elected to a Fellowship at Pembroke. Almost all of G G Stokes' 138 published papers, of which only one was jointly authored, appear in the five volume *Mathematical and Physical Papers* (Cambridge, 1880–1905). The first three volumes were edited by Stokes personally, the last two by his successor in Cambridge, Sir Joseph Larmor*, himself an Irishman. Larmor also published the memoirs and scientific correspondence of G G Stokes in 1907, an important source for later scholars.

Like many other disciplines, mathematical physics has become much more compartmentalized in the twenty-first century than it was in the nineteenth. It is therefore difficult for one modern researcher to offer a comprehensive evaluation of every aspect of Stokes' work. While all his major interests will be mentioned, I hope that readers will forgive me if I use the fields of research with which I am most familiar from my own work to give the flavour of the man.

Stokes' early research was in the area of hydrodynamics, both experimental and theoretical, during which he put forward the concept of 'internal friction' of an incompressible fluid. This work was independent of the publications which were appearing in the French literature at the same time. Stokes' methods could also be applied to other continuous media such as elastic solids. He then turned his attention to oscillatory waves in water, producing the subsequently verified conjecture on the wave of greatest height, which now bears his name.

One of the many places in which the name of Stokes occurs in modern physics is in the Navier–Stokes equations. These fundamental equations for the motion of incompressible fluids were first published in 1822 by the French civil engineer Claude Navier. Modern physicists would find his analysis based on an unacceptable notion of intermolecular forces. Using his own concept of internal friction in fluids, it was Stokes in 1845 who put the derivation of these equations on a firm footing. Thus it is by both names, Navier and Stokes, that these equations are known throughout the world today.

They are used to describe the wake behind a boat or the turbulence behind a modern aircraft, and are employed on a daily basis by aeronautical engineers, ship designers, hydraulic engineers and meteorologists. While simple examples, such as steady flow in a straight channel, can be solved exactly, and some more complicated cases admit an approximate numerical solution by large-scale computer packages, the mathematical problem of the existence and uniqueness of a general solution to the Navier–Stokes equations remains unsolved today.

Such was Stokes' reputation as a promising young man, familiar with the latest Continental literature, that in 1849 he was appointed to the Lucasian Chair of Mathematics. This prestigious professorial chair is currently held by Stephen Hawking. But the chair was relatively poorly endowed, and to augment his income Stokes also taught at the School of Mines in London throughout the 1850s. Besides his links with the School of Mines, he acted, over a period of many years, as consultant to the lensmaker Howard Grubb* who ran a successful and internationally-known optical works in Rathmines. He also acted as advisor on lighthouse illuminants to Trinity House. Stokes' collected works include a paper on a differential equation relating to the breaking of railway bridges, and, following the Tay Bridge disaster, he served on a Board of Trade committee to report on wind pressure on railway structures.

Although appointed to the Lucasian Chair for his outstanding research, Stokes showed a concern in advance of his time for the welfare of his students, stating that he was 'prepared privately to be consulted by and to assist any of the mathematical students of the university'. It is recorded that Babbage, an earlier incumbent, and pioneer of the computer, never once addressed classes. In contrast, Stokes immediately advertised that 'the present professor intends to commence a lecture course in Hydrostatics', which he was still delivering 53 years later, in the last year of his life. Stokes' manuscript notes still exist in the University Library in Cambridge, although his writing was so bad that he eventually became one of the first people in Britain to make regular use of a typewriter. He acquired the first of these, which used only upper case letters, in 1878. The second, used from 1886, also had all capital letters and only the third, used from 1890, had both upper and lower case. Stokes was a prolific correspondent, using the penny post as a modern scientist might use e-mail. He communicated endlessly with the leading scientific figures of his day, most notably Lord Kelvin. Kelvin had enrolled in Cambridge University in 1841, the year in which Stokes had graduated. They became firm friends and researched in similar areas. When Kelvin was appointed professor of natural philosophy in Glasgow in 1846, their correspondence began. This was easily the most extensive correspondence between two Victorian physicists and 656 surviving letters were published in 1990 by David Wilson as a two-volume collection. This is a major source of information about the content and management of Victorian science generally, and interested readers should see David Wilson's 1987 comparative study of these two great physicists (see Further Reading).

Although Stokes occupied a Professorship of Mathematics, he was far from being a pure mathematician. His mathematical results arose mainly from the needs of the physical problems which he and others studied. His paper on periodic series concerned conditions for the expansion of a given function in what we now know as a Fourier series. In the

Stokes in 1892 at the age of 73. Photograph by Mrs F W H Myers. From Stokes 1880–1905 *Mathematical and Physical Papers,* ed J Larmor vols IV and V (Cambridge: Cambridge University Press) vol. V frontispiece.

course of this work he made use of what we now know as the Riemann–Lebesgue lemma some seven years before Riemann. Stokes is also credited with having had the idea of uniform convergence of a series. His major work on the asymptotic expansion of integrals and solutions of differential equations arose from the optical research of G B Airy on caustics behind rainbows, where he was the first to recognize what we know today as Stokes' Phenomenon. He employed the saddle point method for integrals in the complex plane a full decade before Riemann, to whom it is usually attributed. The well-known theorem in vector calculus which bears his name is sadly not due to Stokes, but was communicated to him in a letter by Lord Kelvin, and subsequently set as a problem in the Smith's Prize examination at Cambridge.

Stokes continued his researches in the principles of geodesy (another link with his surveyor great-grandfather), and in the theory of sound, which he treated as a branch of hydrodynamics. But perhaps his major advance was in the wave theory of light, by then well established at Cambridge, examining mathematically the properties of the 'luminiferous ether', which he treated as a sensibly incompressible elastic medium. 'Ether' was the old-fashioned name given to the medium filling all space, which was thought to carry light waves as vibrations analogous to sound waves. It predated the theory of light as an electro-magnetic phenomenon introduced by the Scottish physicist James Clerk Maxwell, who was appointed to the Cavendish chair of experimental physics at Cambridge in 1871. After Hertz's 1887 experiment showed that electromagnetic waves could be generated by an electric circuit, the concept of the ether was attacked by another Irish physicist and correspondent of Stokes, George Francis Fitzgerald*, the leading follower of Maxwell, Erasmus Smith Professor of Natural and Experimental Philosophy at Trinity College Dublin.

The concept of the ether enabled Stokes to obtain major results on the mathematical theory of diffraction, which he confirmed by experiment, on polarization of light and on fluorescence, which led him into the field of spectrum analysis. The spectrum of light passed through a prism gives us the characteristics of its source; the process of examining quantitatively this spectrum is known as spectroscopy. When light is shone on or through a material, its spectrum is modified because the material is absorbing light of certain preferred wavelengths. In some cases it may even happen that a part of the absorbed light is radiated again at a different wavelength. This is called fluorescence, whose discovery is attributed to Stokes. His last major paper on light was his study of the dynamical theory of double refraction, presented in 1862, although late in life, from 1896 onwards, he was involved in the early investigation of Röntgen rays, now known as X-rays.

The second half of Stokes' life was increasingly taken up with scientific and academic administration. A major reason for this change was that in 1851 he had been elected a Fellow of the Royal Society, and shortly afterwards, in 1854, became Secretary of the Society, where he performed an important role in advising authors of research papers of possible improvements and related work. Many famous scientists tried out their half-formed ideas on Stokes, who was also extremely active in the British Association for the Advancement of Science.

His close colleagues regretted his taking on these administrative duties and P G Tait even went so far as to write a letter to *Nature* protesting at "the spectacle of a genius like that of Stokes wasted on drudgery [and] exhausting labour". His friend Kelvin wrote to him in 1859 of "the importance to science of getting you out of London and Cambridge, those great juggernauts under which so much potential energy for original investigation is crushed". As early as 1849, Kelvin had proposed that Stokes should join him in Glasgow as Professor of Mathematics. Stokes, a man of strong principles, had refused to apply because, as a lifelong Anglican, he would have had to make a statement of conformity to the established (Presbyterian) Church of Scotland. But even after the religious tests had been removed, Stokes did not apply for the vacant Professorship of Astronomy in 1859.

In 1859 Stokes vacated his Fellowship at Pembroke, as he was compelled to do by the regulations at that time, on his marriage to Mary Susannah, daughter of Dr Thomas Romney Robinson, FRS, Astronomer at Armagh. Following a change in regulations, he was subsequently able to resume his Fellowship, and for the last year of his life served as Master of Pembroke. Shortly after their marriage the couple moved to Lensfield Cottage, a happy and charming home, in which Stokes had a 'simple study', and conducted experiments 'in a narrow passage behind the pantry, with simple and homely apparatus'.

Prior to their marriage Stokes, who, as we have already remarked, was a tireless writer of letters, had carried on an extensive (one letter ran to 55 pages) and frank correspondence with his fiancée. In one letter, the theme of which will be familiar to all spouses of research physicists, he states that he has been up until 3 a.m. wrestling with a mathematical problem, and fears that she will not permit this after their marriage! Based on other remarks in this highly personal correspondence, David Wilson, in his 1987 book on Kelvin and Stokes, suggests that "Stokes himself may have welcomed what others regretted—his abandonment of the lonely rigours of mathematical physics for domestic life and the collegiality of scientific administration".

Another result of his marriage was a rekindling of his interest in water waves. Stokes and his family visited his father-in-law Dr Robinson every summer, and took regular holidays at resorts on the north coast, most frequently at Portstewart, from whence they made excursions to the Giant's Causeway. Although primarily a theoretician, Stokes was not afraid to experiment: he measured the waves breaking in the Land Cave there, and also on the sloping sandy beaches at Portstewart.

Typical of his letters to Kelvin about this time is one written from the Observatory, Armagh, on 15 September 1880. Stokes had made a mathematical conjecture that the angle at the crest of the wave of greatest height should be 120°. "You ask if I have done anything more about the greatest possible wave. I cannot say that I have, at least anything to mention mathematically. For it is not a very mathematical process taking off my shoes and stockings, tucking my trousers as high as I could, and wading out into the sea to get in a line with the crests of some small waves that were breaking on a sandy beach.... I feel pretty well satisfied that the limiting form is one presenting an edge of 120 degrees."

He writes on the same topic a week later and Kelvin replies from his yacht, the *Lalla Rookh* at sea in the Clyde. In another letter Kelvin writes "Will you not come and have a sail with us and see and *feel* waves? We would take you away out to the west of Scilly for a day or two if that would suit best." Stokes replies light-heartedly: "It is not easy to say where to find a man who owns a yacht, but I write on spec, and at any rate you will soon I suppose be back in Glasgow."

Despite all this seaside holiday paddling, Stokes' interest in ocean waves was a serious one, undertaken in consequence of his membership of the Meteorological Council. He was aware that long waves radiating from distant storms travelled faster than short waves from the same source. In situations where unusually high seas were observed in the absence of a local wind, Stokes was able to analyse records of the direction and period of the waves, to predict the location and direction of travel of the storm which had given birth to them. Larmor's 1907 Memoir contains his fascinating correspondence with the Admiralty Experimental Station at Torquay, and various sea captains, most notably Captain William Watson of s.s. *Algeria*, on these observations from ships' logs. Stokes also advised on meteorological instruments, and a sunlight recorder designed by him was in use in the station at Valentia until recent times.

Stokes was as selfless in public life as in his professional life. He served as a Member of Parliament for Cambridge University from 1887 to 1892, overlapping with his Presidency of the Royal Society (1885–1890). It was a time of conflict in British science between the supporters of Creation as set out in the Bible, and followers of the Theory of Evolution. The Anglican Stokes and the Presbyterian Kelvin were firmly in the creation camp, while the Irishman John Tyndall and Thomas Henry Huxley were leading supporters of evolution within the Royal Society. Huxley perceived a conflict of interest and wrote anonymously in *Nature* that, as President of the Royal Society, Stokes should not simultaneously be a Member of Parliament.

A deeply religious man, Stokes had always been interested in the relationship between science and religion. From 1886 to 1903 he was President of the Victoria Institute, whose aims were 'To examine, from the point of view of science, such questions as may have arisen from an apparent conflict between scientific results and religious truths; to enquire whether the scientific results are or are not well founded'.

Many honours were bestowed on him in later life. He was made a baronet (Sir George Gabriel Stokes) by Queen Victoria in 1889, was awarded the Copley Medal of the Royal Society in 1893, and in 1899 given a Professorial Jubilee (50 years as Lucasian Professor) by the University of Cambridge. Stokes died at Lensfield Cottage at 1 a.m. on Sunday 1 February 1903.

Although obsessed with his scientific work, Stokes had excited feelings of warmth and admiration among his contemporaries. Some colleagues felt that he could have done more in the field of physics in later life, but Stokes himself seemed to find fulfilment in his role as a senior figure in the Victorian scientific establishment. Lord Rayleigh and Lord Kelvin both published obituaries of Stokes, Kelvin observing at the time that his heart was in the grave with Stokes. In its leader of 3 February the *Times* wrote "Sir G Stokes was remarkable... for his freedom from all personal ambitions and petty jealousies.... It is sometimes supposed—and instances in point may sometimes be adduced—that minds conversant with the higher mathematics are unfit to deal with the ordinary affairs of life. Sir George Stokes was a living proof that if the mathematician is only big enough, his intellect will handle practical questions so easily and as well as mathematical formulas.''

Further Reading

(*All have been used in the preparation of this short account.*)

G G Stokes and his Precursors on Water Wave Theory, A D D Craik, lecture delivered at the Stokes Summer School, Skreen, County Sligo, August 2000 (to be published).

Creators of Mathematics: the Irish Connection, Ken Houston (ed) (University College Dublin Press, 2000) chapters 5, 8.

Memoir and Scientific Correspondence of the Late Sir George Gabriel Stokes, Sir Joseph Larmor (ed) (Cambridge University Press, 1907).

'The Mathematical Work of G.G. Stokes', R B Paris, *Mathematics Today* **32**(3/4) (1996) 43–46.

Mathematical and Physical Papers Volumes I–V, Sir G G Stokes (Cambridge University Press, 1880–1905).

'Rich Silence: Some of Stokes' Contributions to Physics', D Weaire, *Mathematics Today* **32**(5/6) (1996) 81–83.

Kelvin and Stokes, a Comparative Study in Victorian Physics, David B Wilson (Adam Hilger, Bristol, 1987).

The Correspondence between Sir George Gabriel Stokes and Sir William Thomson Baron Kelvin of Largs, David B Wilson (ed) (Cambridge University Press, 1990) (2 volumes).

'George Gabriel Stokes 1819–1903, an Irish Mathematical Physicist', Alastair D Wood, *Irish Mathematical Society Bulletin* **35** (1995) 49–58.

Useful Websites

http://www-history.mcs.standrews.ac.uk/history/References/Stokes.html
http://webpages.dcu.ie/~wooda/Stokes.html

John Tyndall

1820–1893

Norman D McMillan

John Tyndall was born, probably in 1820, in Leighlinbridge, County Carlow, the son of an RIC officer and land agent for Steuart, the local Protestant landlord. His father, who came from a family of small landowners in County Kilkenny, supplemented his meagre income by working as a boot-mender. The family moved from Carlow to Castlebellingham, County Louth, for a short period, and then, on the family's return to Leighlinbridge, John completed his education under the tutelage of John Conwill. Despite the fierce protests of his Protestant neighbours, John's father insisted that 'even if he was taught on the steps of the altar' he would send his son to Conwill, who was a teacher of some renown. John left Conwill at the very advanced age of 19 years, equipped, as his father had hoped, with a well-rounded vocational training. This education included English, logic, book-keeping, drawing and, most importantly, surveying and associated mathematics.

He joined the Ordnance Survey Office in Carlow from school, moving from there to Youghal, County Cork. Here he began his lifetime efforts at private study. Tyndall, in his personal journal, showed that he had no sympathy for the declining Gaelic language in use around Youghal, and one feels, from reading his own thoughts, that he was by then anxious to leave his native shores. Tyndall consequently felt fortunate to be chosen, in 1842, to transfer to the English Survey where he was posted in Preston, Lancashire.

However, his discontent with the exploitation of Irish workers in the Survey culminated in a formal protest, and, as leader of the malcontents, Tyndall was summarily dismissed in November 1843. Tyndall thereby became a *cause célèbre* in Lancashire, and wrote a series of articles protesting against the actions of Robert Peel in the radical *Liverpool Mercury.* When a petition to the Master General of Ordnance failed, Tyndall returned home. He spent a period of nearly a year from 1843 back in Leighlinbridge, where he read Thomas Carlyle's *Past and Present* avidly. (Tyndall and Carlyle were later to become close friends.)

Tyndall eventually found work as a surveyor with Manchester partners Nevin and Lawton. He left to work on the West Yorkshire

Line as a railway surveyor at the height of the railway fever. Operating for much of this time from a base in Halifax, he was, however, employed on lines as far south as Bedfordshire. It was during this time that he developed his prodigious walking capacity and developed the stamina that was to serve him so well later as a mountaineer. In addition, he gave evidence before Parliamentary commissions on competition for new rights, finding this a valuable training ground and introduction to high-powered metropolitan affairs.

Tyndall's interest in physics began in 1847 when he met George Edmondson, who at the time was endeavouring to introduce the elements of experimental science into his Quaker School, Tulketh Hall, which was in Preston. Tyndall helped him with this project by doing some teaching. Shortly thereafter, Edmondson entered into negotiations with representatives of Robert Owen to take over his last great communist social experiment known as 'Harmony Hall' near Stockbridge in rural Hampshire. This amazing educational facility was owned by the then financially hard-pressed Owen. It was unique in that it stood on its own large 500 acre self-sufficient farm, boasting truly palatial buildings, with the latest and most lavish educational facilities ever erected. The educational facilities supporting the agricultural school included a purpose-built science laboratory, printing office, and carpenter's and blacksmith's shop.

When the Quakers took over the school, for propriety they changed the name to Queenwood College. Edward Frankland, another passionate self-improver, was appointed superintendent of the chemistry laboratory, and the younger Tyndall, thoroughly in the thrall of the older man, was appointed to teach mathematics and to supervise the engineering department. They immediately began a programme to teach self-study. The two worked in tandem in a remarkable year of effort, to establish the first-ever programme of practical science and engineering in Great Britain and Ireland, in what was almost certainly the first-ever school laboratory. Their innovation marks out a landmark in English education, which was developed on a national arena with the appointment in 1853 of both men as examiners in the Science and Arts Department under the supervision of Lyon Playfair.

Frankland was the first examiner in chemistry and Tyndall in physics. Physics had, in fact, developed as an entity in revolutionary France, and came to Britain via Germany. Playfair and Tyndall defined, through this school examination, and thereby the curriculum, the exact shape of the school subject of physics, and, as a consequence, fundamentally shaped the British profession of physics. Tyndall was also appointed in 1857 as the first Chief Examiner for military examinations for the Royal Engineers and Royal Artillery.

In October 1848 Frankland and Tyndall left Queenwood to attend—at their own expense—Marburg University, then the centre of the radical

Tyndall in 1850 (courtesy of the Royal Institution).

scientific materialist movement. Tyndall produced a PhD in 1849 after a two-year period of intense study. Two subsidiary subjects—chemistry and physics—supplemented this geometric study of screw surfaces. He studied chemistry under the illustrious Robert Bunsen, whose own research interests in spectroscopy were to fire and shape much of Tyndall's own later research ambition.

On completing his studies, he moved decisively towards physics, collaborating with Heinrich Knoblauch, who had arrived in Marburg from Berlin, to produce his first published paper on what was then the hot new research topic of diamagnetism. This field was to be his major research interest for six years. A loan from his best friend Thomas Hirst made it possible for him to extend his stay in Germany long enough to produce a second memoir with Knoblauch, and then to spend several months in Gustav Magnus's laboratory in Berlin. Here he cemented what were to be lifelong ties to the elite of Germany's scientific community, including Magnus, Poggendorff, Heinrich Dove, Emil de Bois-Reymond and Rudolf Clausius.

In June 1851, he returned to Queenwood and sought positions in Toronto, Sydney, Cork and Galway, without success. He continued his work on diamagnetism with small equipment grants he obtained from the Royal Society with help from fellow-Irishman Edward Sabine and

Faraday. His considerable work on this topic was only much later consolidated into a book published in 1870 under the title '*Researches on Diamagnetism and Magne-crystallic Action*' By this time the results were not of current interest.

Tyndall gave the Discourse at the Royal Institution (RI) of Great Britain on 11 February 1853 on the topic of diamagnetism, presenting his own theory opposing that of Michael Faraday. He was invited to give a second Discourse and a course of lectures and, by May 1853, he was elected to the Chair of Natural Philosophy at the RI. Tyndall's friendship with Faraday developed despite their being very different characters; Tyndall repaid this friendship on Faraday's death, with his excellent scientific biographical study *Faraday, as a Discoverer* published in 1868. Tyndall indeed abandoned his earlier German methodology, and became an open disciple and advocate of Faraday's pure experimental method.

Tyndall began his first series of independent researches on slatey cleavage, the real objective of which was to experimentally support Faraday's geological theories. This work, however, brought him into conflict with James D Forbes of Edinburgh, and thus almost the entire Scottish establishment, including the formidable duo, William Thomson* (Lord Kelvin) in Glasgow, and P G Tait in Edinburgh. It is now generally agreed that Forbes won the debate on glacial motion over the theory first advanced by Bishop Rendu and subsequently supported by Faraday. Tyndall, however, vigorously defended Faraday's theory of regelation of ice in public lectures to celebrity audiences at the RI. His persuasive lectures were always enlivened with new and striking experimental demonstrations, such as the one devised in this dispute.

He presented his crucial experimental demonstration, in which a single solid block of clear ice was produced by compressing pieces of ice in a closed tube by hammer blows. He progressively honed his natural gifts of delivery with the anticipation of making one of his frequent controversial statements, which must have enlivened any audience. Tyndall became, by common assent, the greatest of all expositors of science of his day.

From the outset, Faraday's experimental method, despite his own personal acclaim and international standing, was a serious threat to the mathematical domination of what have been called the 'Gentlemen of Science'. The latter were an ascendant latitudinarian tendency in the established church and universities. These 'mathematical power-brokers in science' very effectively controlled science and all its institutions at that time. They profoundly mistrusted experimentalists and especially the new professional class of scientists who were wedded to this ideology.

Furthermore, Tyndall and Frankland organized a group of eight London-based friends from their base on Albermarle Street, in 1864.

Left to right: Michael Faraday, Thomas Henry Huxley, Charles Wheatstone, David Brewster, John Tyndall (courtesy of AIP Emilio Segrè Visual Archives, Zeleny Collection).

The ninth member, William Spottiswoode, was admitted at the second meeting, and the club eventually became known as the X-Club. It was so named because of the nine members, plus their acknowledged, but always absent, Xth member and leader, Charles Darwin, who was too retiring either to attend X-Club meetings, or to publicly defend his own theory. The X-Club members all adopted epithets, viz. Xccentric, Xalted, Xpert, Xperienced, Xqisite, Xemplary, Xhaustive, Xtravagant and Xcellent respectively for Tyndall, Huxley, Frankland, Joseph Hooker, John Lubbock, George Busk, Herbert Spencer, T A Hirst and Spottiswoode.

These professionals had taken onto themselves the role of spokesmen for the theory of evolution after the 1859 watershed of the publication of the Darwin's *On the Origin of Species*. The Xs' activities greatly diminished in the public's eyes the gravitas of the 'Gentlemen's' pronouncements as arbiters of scientific truth delivered at large public gatherings of the British Association or elsewhere. The Xs met monthly before meetings of the Royal Society for some 30 years and, no doubt, coordinated their efforts in influencing election results and other matters inside this recently reformed British scientific governing body. It was popularly believed in later years that the Xs ran science.

Tyndall's own research from 1860 on radiation through gases and vapours had provided much of the essential basis for the science of meteorology, which was absolutely central to the scientific arguments on life on earth. His book *Contributions to Molecular Physics in the Domain of Radiant Heat* appeared in 1872, marshalling an array of new facts supporting the theory of evolution. He had come upon this virgin field in 1860 through an invention. Realizing that galvanometers were very insensitive instruments for directly measuring the amount of infrared flux from a Melloni thermopile, he devised a new system to measure in a differential mode.

He invented the double-beam null balance detector, which was the first optoelectronic detector circuit, and which was in practice hundreds of times more sensitive than any previously used. Tyndall's circuit initially balanced signals from a reference beam with those from the measurement beam, the gas or vapour under study being either pumped out of, or alternatively added to, the experimental tube. As a consequence, he could measure both the absorptions and emissions of gases and vapours for the first time, from the unbalanced signal produced, for example, by evacuation.

He should be considered the founder of optoelectronics, and a father of both infrared analysis and spectroscopy, based on his subsequent exhaustive heat studies over the next decade. Remarkably, this work formed the basis of three Royal Society Baconian lectures, in which Tyndall formulated a quantitative understanding of atmospheric physics. Tyndall opened up the debate on the greenhouse effect, which continues to this day, and used his experimental knowledge on the large absorptive capacity of water vapour to explain, for instance, meteorological conditions in deserts and other climates to great effect.

In the latter part of these studies he turned to shorter wavelengths and found that these rays of 'high refrangibility' (ultra-violet) caused photochemical reactions. The resulting clouds of small particles scattered visible light to produce colours and in particular the vivid blue of the sky, known frequently as 'Tyndall blue'. Thus Tyndall was the first scientist to explain why the sky is blue. His investigations into scattering have led him to be honoured by the naming of 'Tyndall scattering' for scattering from particulate matter.

Tyndall, made an impressive and comprehensive experimental study of the phenomenon, including polarization studies. He pointed out with amazing prescience that space would be black, and in 1869 produced a cometary theory to explain the tails of comets. Tyndall provided in his monumental series of experiments instruments for measuring radiation in gases and vapours, for observing light scattering, fluorescence and photochemistry in gases, vapours and liquids. Here we find the experimental basis of environmental scientific monitoring.

Tyndall openly and consistently fought for an atomicist and progressively reductionist description of matter. He played a central role in the establishment of the appropriately named evolutionist journal *Nature*, which would again have been deeply resented and seen as threatening by the scientific establishment, with their own established journals that gave them control of published science. He took the lead in Britain to extend the battle to debunk Kant's 'things-in-themselves'. Indeed, he had gone much further, tackling religion head-on, with his criticism of religious revelation and the efficacy of prayer, in the long debate from 1861 to the mid-1870s. He suggested a 'prayer gauge or test' and openly stated his conviction that theological speculations were harmful.

Tyndall had made formal his break with religion in aiding Huxley in establishing a new-world view of the sceptical philosophical materialistic position of 'humanism'. Thus, he, with others such as Huxley and Spencer, had a fervour to imbue new leaders of the nation with genuine knowledge of the laws of nature and society. In the last analysis, however, these men, like their German counterparts, demanded a new scientific and progressive ideology for the nation in a new imperial age. This was indeed the 'Age of the Empire', which demanded scientific solutions to new pressing competitive nationalistic demands.

In philosophy, Tyndall had gone well beyond his co-thinker Huxley, taking an enormous leap in developing the position of the German reductionists. In his 1870 lecture for the British Society for the Advancement of Science, titled *On the Scientific Use of Imagination*, he had notably presaged de Bois-Reymond's 1872 lecture that is seen, probably wrongly, as 'the' philosophical milestone on this issue. He extended substantially his previous position, given in his BAAS lecture at Norwich, on the role of the imagination in science, as *an active force of matter*. He had thereby included human volition into an atomicist-evolutionary philosophical description of matter. He flatly stated that, from a fundamentalist evolutionary position, all philosophy, science and art are potential in the fires of the sun.

He had argued for a Fichtian (modified Kantian) position in a *world that is*, which is fundamentally separate from *the world that might be*. Thus

Tyndall in 1865 (courtesy of the Royal Institution).

he strategically sought to exclude science from discussing the existence of God, by accepting the Kantian domination of God in the world outside of science. This placed a prohibition on religion having any say whatsoever in science, which was exactly, despite his protestations, what he was aiming to do. In the Belfast Address he 'rejected the notion of a creative power', and referred the 'choicest material of the teleologists' to natural causes.

The Belfast Address was pure, undistilled, revolutionary, anti-religious, theorizing from the perspective of organized religion. When the story of the Belfast Lecture appeared on the front pages of papers throughout Europe and North America, Tyndall compounded the situation of alarm by adamantly refusing to step back from the hornets' nest of controversy he had raised. Rather, he stoked up the controversy with his *Apology for the Belfast Address*, which was not an apology at all, but rather a forthright defence of his position set out in Belfast.

Churches all over the world immediately denounced Tyndall's atheism and in Ireland a Catholic Church Pastoral letter was issued in 1875, written for Primate Cullen, it is confidently believed, by Patrick Francis Moran, who was a contemporary of Tyndall from Leighlinbridge. Moran, ironically, now has a commemorative stone standing alongside that of Tyndall's in a field in their native village, at a time when the

Catholic Church has reconciled itself to evolution. This Pastoral Letter stated "that under the name of science", Tyndall obtruded blasphemy upon the Catholic nation, and warned in the very strongest terms against those who might follow Tyndall down the evolutionary road.

When lecturing, Tyndall never read from his carefully prepared notes, but rather put himself *en rapport* with his audiences. His lectures were described as the fairyland of science, in that the poetic and imaginative aspects of science were marshalled as a means of education. His lectures to children were especially well received and for many years he delivered the Royal Institution's famous Christmas Lecture series. These lecture notes in effect grew into the series of school textbooks that he had reluctantly declined to write for the new Department of Science and Arts' physics examination programmes.

He published eleven popular physics textbooks based around his lectures on Sound (1867 and 1873), Light (1869, 1873 and 1878), Electricity (1870, 1876 and 1881), Heat (1863 and 1877) and Ice, Water, Vapour and Air (1871). Most of these books ran to edition after edition, and were also printed abroad. There can be no doubt that these books most profoundly shaped the practice of school physics courses in their first formative 50 years, and many schoolteachers still covet copies of Tyndall's books today. Tyndall himself naturally took an active interest in the training of school science teachers. He was involved in the first ever Science and Arts Department under the auspices of a committee of the Council for Education at South Kensington Museum in 1861. His lecture *On Experimental Physics* was published in the same year. He regularly interacted with schoolteachers, for whom he was an exemplar, and who came to the RI to see his lectures.

The other great innovation made by Tyndall in his lectures was that he devised inventions specifically for the lecture. For example, he gave the first public demonstration, given many times subsequently, of the guided light-pipe, which in our own time has led to the fibre optic. This invention was made in 1854, not as quoted in most fibre optics books on dates such as 1877. The demonstration came as a final show-stopping finale in an RI lecture entitled *Phenomenon Related to the Motion of Liquid*. He had earlier published two papers on water jets and motion of liquids in 1851, which prepared the ground for his light pipe discovery. A little known fact is that in his classic textbook *Heat a Mode of Motion* he devised an experiment of a hollow polished pipe guiding an infrared wave, which, of course, marks the discovery of the waveguide.

A second major lecture coup followed in 1854 when he resolved experimentally the disputed issue of geyser action, by building an artificial geyser; he presented his findings in an RI Discourse.

In 1871, prompted by seeing a dead, but otherwise totally unmarked woman being taken from a building where she had expired from suffocation, he devised an improved respirator. The firemen's existing cotton-wool respirators being useless against resinous smoke, they were unable to save the woman. In his RI Discourse, he gave details of a helmet respirator employing a new filter. This multi-layered filter had a cotton pad moistened with glycerine, then the air was passed through a thin layer of dry cotton-wool to a layer of charcoal to adsorb the hydrocarbons, and finally through a second layer of dry cotton wool. Air was exhaled through a separate exhaust valve to avoid wetting the various filter layers by breath. This is the forerunner of all modern respirators.

Amongst his most important scientific inventions was his famous meteorological equipment. He created an artificial sky and sunset with photochemical reactions, and he devised a cloud chamber (which was developed by C T R Wilson in his seminal studies in Cambridge, for which he was awarded the Nobel Prize in Physics).

The very limited knowledge of medical practitioners at the time was evident to Tyndall in other ways. The biochemist Louis Pasteur was having great difficulty in dealing with the attacks on his germ theory by the entrenched and hostile French medical establishment. Tyndall was induced by Pasteur himself to begin working on disproving the theory of 'spontaneous generation' and hence settle the on-going fierce dispute in favour of the germ theory. Their extensive correspondence stretches from 1871 and shows that Tyndall was by no means the junior partner in their collaboration.

Tyndall's entry to the field of disease and infection was through studies on dust and disease using his nephelometric technique in a paper entitled *Dust and Disease* published in 1871. Abiogenesis, or spontaneous generation, held that living organisms could be formed from inorganic matter. Tyndall, after incredible commitments of time over nearly a full decade, succeeded in producing methods of rigorous sterilization, and was thus able to prove the germ theory by producing reproducible bacteriological experiments on sterilization. Pasteur himself could not achieve this vital experimental reproducibility, because his process of sterilization, known today as pasteurization, had just a single heat stage, and it consequently does not kill the spores of bacteria. The bacteria die but the spores then allow the bacteria to recolonize the medium.

Tyndall showed in 1877 that the only reliable method of sterilization, known as Tyndallization, is a process of intermittent heating. These repeated processes produce sterile media, because they kill the bacteria, but then also the spores of the bacteria. Tyndall's use of light scattering to ensure that his apparatus was clear of airborne contamination was a further reason for success, along with his mastery of filtration methods which excluded airborne contamination. His monumental researches on the topic were published in a book entitled *Essay on Floating Matter of the Air in Relation to Putrefaction and Disease* and published in 1881. This work clearly establishes Tyndall's claim to be the founder of the experimental science of bacteriology, working on well-established Baconian principles. Pasteur, on the other hand, stands as the theoretical founder of bacteriology and microbiology. Both Huxley and Tyndall saw this scientific battle as their final frontier, being an essential fundamental defence of the theory of evolution, which of course required evolutionary processes ruling out any possibility of spontaneous generation.

Further Reading

John Tyndall: Essays on a Natural Philosopher, W H Brock, N McMillan and C Mollan (eds) (Royal Dublin Society, 1981). A 250th anniversary publication of the Society.

Tyndall the 'X'emplar of Scientific and Technological Education, N McMillan and J Meehan (NCEA, Dublin, 1981).

John Tyndall, Norman McMillan in: *Dictionnaire des Philosophes de France* (Paris 1982; and enlarged entry in 2nd edition, 1993).

'British Physics—The Irish Role in the Origin, the Differentiation and Organization of a Profession', Norman McMillan, *Physics Education* **23** (1988) 272–278.

'John Tyndall and the Foundation of the Sciences of Infra-Red Spectroscopy and Nephelometry', N D McMillan and L Vallely, *Technology Ireland* **21**(4) (1989) 37–40.

Two articles by Norman McMillan written for the people of Carlow were published in the *Kilkenny People*: 'John Tyndall: Shining Beacon to Struggling Youth' (15 and 22 April 1977), and 'John Tyndall—Remembered Everywhere but at Home' (30 September 1977).

Samuel Haughton

1821–1897

Norman D McMillan

Samuel Haughton came from a very industrious and prosperous Quaker family involved in malting and milling in Carlow. The Haughtons had large commercial properties alongside both banks of the rivers in the town. Samuel was born on 21 December 1821 to Sarah Handcock, whose father was a successful linen merchant in Lisburn. His family background was consequently one of some considerable local celebrity. Samuel was from the earliest age trained as a naturalist, with the benefit of long nature walks on the side of the River Barrow and the adjacent bog lands, in the company of his schoolmaster. The pair also apparently ventured into the neighbouring hills, where the boy acquired a lifetime's interest in geology.

At seventeen, Samuel entered Trinity College Dublin, and rapidly made his mark 'being very quick of apprehension and with a tenacious memory'. The university was at this time at its scientific zenith, boasting several academics with international reputations, including William Rowan Hamilton* (1805–1865), Humphrey Lloyd* (1800–1881) and, of special importance to Haughton's subsequent career, James MacCullagh* (1809–1847). Amongst his own generation of students were included the notable mathematicians and teachers, George Salmon (1819–1904), John Hewitt Jellett (1817–1888), James Booth (1806-1878), and, again one of special importance to Haughton's career, the radical, Joseph Allen Galbraith (1818–1891).

Trinity had the most competitive undergraduate and fellowship examinations in Britain. Haughton excelled in this environment; he obtained the Lloyd Exhibition in 1842, and went on to graduate with a Moderatorship in 1844. A mere seven months after graduating, in 1845, he was successful in the Fellowship examinations, in direct competition with some very illustrious names. Fellowship of Trinity was the supreme academic achievement in Dublin, and Haughton became the youngest ever Fellow.

He began his postgraduate research career sharing rooms with, and in effect studying under, MacCullagh. He was applying mathematical methods in physics and chemistry, and in 1848 was awarded the prestigious Cunningham Medal of the Royal Irish Academy for only his second independent publication. However, this first phase of Haughton's varied career came to a premature end. Tragically MacCullagh committed suicide in 1847. In any case, Haughton may have been attracted by the prospect of the Chair of Geology, which was soon to be vacated, and his researches moved in this general direction.

Haughton began his academic career working in the then new Engineering School. The very existence of this School marked a significant University reform. It was the first in any of the old established universities, and had been brought into existence by the long and determined agitation of Humphrey Lloyd, MacCullagh and other reformers. The backdrop to this educational reform was the Royal Commission of 1851. Fortunately the Commissioners included no less than six Trinity men, and the Commission report was duly impressed with the evidence of young reformers such as Haughton, and the example they provided of Trinity 'self-reform'. Haughton's youthful brilliance and industry were then duly rewarded in 1851 by his election to the Chair of Geology in this new School.

There existed no model of what subject material should be taught in engineering courses, a situation that gave Haughton, working with his young friend Galbraith, an exceptional opportunity. The pair pioneered an important set of courses and textbooks for a wide range of new subject areas. The importance of the *Galbraith and Haughton Scientific Manuals* (hereafter referred to as *Manuals*) was greatly increased because examinations had been introduced in 1848 for the Indian Civil Service and for Commissions in the Artillery and Royal Engineers. Courses were run in Trinity for these new public examinations, and the *Manuals* written for Trinity students were used widely in other British institutions training men for the Empire.

These *Manuals* included at least 19 separate books. Details given below are certainly not complete, despite serious searching of the archives, but the *Manuals* did include the following: *Elementary Mathematics* (1851), *Plane Trigonometry* (1851), *Arithmetic* (2nd edition, 1855), *Mechanics* (1854), *Optics* (1854), *Hydrostatics* (1854), *Astronomy* (1855), *Euclid Books I and II* (1856), *Experimental and Natural Science Series* (4 volumes 1859-65), *Euclid Book I to IV* (1859 to 1863 in 2 volumes), *Algebra* (1860), *Tides and Currents* (2nd edition, 1862), *Steam Engines* (1864), *Geology* (1865) and *Mathematical Tables* (4th edition, 1887). The *Manuals* were published all over the world; editions from London, Paris, New York, Bombay and Sydney have been found. In many cases they appeared in many editions; for example, the *Hydrostatics Manual* had its ninth edition in 1891.

The importance of these books should not be underestimated, as at this time there were few textbooks for these subjects and the ones that were available were out of date, and in any case not well written for examination students. The *Manuals* were perfect for their task; being styled on the classic Trinity 'cramming' textbooks, they were very well tailored for both the pocket and the limited examination scope of interest of their target readership.

As well as the *Manuals*, Haughton published an important book, *On the Rotation of a Solid Body round a Fixed Point*, which was an account of MacCullagh's lectures on the subject, and was written shortly after his suicide. Much later, in 1880, together with J H Jellet, he edited an edition of MacCullagh's *Collected Works*. He also wrote and edited 11 other educational books, and produced a very widely acclaimed book, *Six Lectures on Physical Geography* (1880), which had considerable influence, not least on his protégé, John Joly* (1857–1933), who was to develop this field further.

Haughton devoted a decade to researches to geology, and made many notable contributions, including, in particular, some lasting ones in physical geology. He had acquired from his training with MacCullagh a mastery of the optical properties of crystals, and could have considered himself, at the time of his election to the chair, to be at the very forefront of mathematical modelling in geophysics, with his work on the earth and its inner structure, and his calculations on the depth of the sea.

A hot area in the geological research of the day concerned physical mechanisms of geological formation, and Haughton rapidly moved into this area in his work on tides. His geological researches eventually, and quite typically for Haughton, stretched over many fields including regional geography, stratigraphy, palaeontology, mineralogy, petrology, structural geology and economic geology (mining). His use of chemical rock analysis was an innovation, and was an important contribution to geochemistry. His use of distortion of fossils to measure the development of rocks has remained significant to this day.

His later work in geology was distinguished by his extensive development of physical geology, and in his *Notes on Physical Geology* of 1877 to 1881, he drew again on his early physics training. He began consideration of changes in the earth's axis, changes in the earth's orbit, and the age of continents, in important contributions to the 'Age of the Earth' controversy. In this dispute, he was a firm ally of Kelvin*. From his first publications in 1857 on Arctic fossils, whenever possible he used his results to support Kelvin's estimate for the geological age of the earth of 100 million years. In 1861, Phillips, who had briefly held the chair of geology in Trinity, had independently estimated from sedimentary stratification, that the age of the earth was 96 million years.

Haughton carried forward another estimate, based on the existence of fossils in the Arctic, which he believed were thus situated in these icy climes because of the gradual cooling of the earth. His estimate of 2298 million years for the period between the time of the formation of the oceans and the beginning of the tertiary period was clearly too big from his own anti-evolutionist perspective. He published the result, however, which gave support to the uniformitarians. In 1878, he reworked the calculation using Rossetti's law of cooling. He assigned a temperature of the Polar Regions for each period, as the earth cooled from the time that water began to condense at 212°F to its present average temperature, which is near 0°F, but his results were puzzling.

These estimates required neither vast expansion of post-Miocene time, nor the contraction of the Paleozoic, Mesozoic and early Cenozoic eras. He employed in this stratigraphical calculation the widely accepted rate of continental denudation using Geike's estimate of one foot per 3000 years, and assumed stratified rock to possess a thickness of 177 200 feet. He then unfortunately used dubious reduction techniques to contrive a result of 152 million years, that was compatible with Kelvin's estimate for pre-Miocene times, but significantly down from the uncomfortably large period of 1.5 billion years.

In his final major contribution to this enormously important debate, he collaborated with J Emerson Reynolds (1844–1920), Professor of Chemistry in Trinity. Their work on the drag of water and air on air was an effort to improve the physical basis for the estimate of the tidal friction calculation. This work in turn inspired his colleague, George Howard Darwin (1845–1912), to begin work on the problem. From a hypothetical model of the genesis of the earth–moon system, and applying essentially the Haughton tidal theory, Darwin came up with an estimate of 54 million years for the formation of our earth–moon system, with the existing duration of months and days. This astronomical estimate was conveniently in agreement with Kelvin's figure, and the paper received much notice from the contemporary scientific fraternity. The acclaim accorded to Darwin's estimate, however, robbed Haughton of much of his richly deserved priority for initiating the work.

Haughton's related researches published in 1880 on the annual discharge of large rivers were an attempt to improve Phillips' sedimentation calculation of the age of the earth. Late in life, Haughton's mature reflection on the geological age of the earth edged up to 200 million years. His varied approaches to chronology of the earth were a significant and many-sided contribution to an important scientific debate.

Haughton's reliance on physical principles unfortunately and paradoxically served to limit greatly his immediate historical legacy, but interestingly it is worth noting as a postscript that it was his early microscopical geological researches in 1858 on the biotite from Ballyellen, County Carlow, that resulted in the discovery of pleochroic haloes. These small spherical features in rock samples are found in some radioactive rocks and result, as shown by Joly in 1907, from the presence of a small quantity of either uranium or thorium emitting alpha particles. The fixed range of the alpha particles that are emitted randomly in all directions produces spherical features. Radioactivity was of course the source of heat energy to the earth that remained unknown to Kelvin and Haughton. This energy was subsequently used by Joly and others to extend Kelvin's estimate of the age of the earth.

Haughton produced a total of 81 papers on diverse topics of geology and physical geography, and five on Arctic travel which arose out of his research interest in Arctic fossils. Haughton's work in geology in its totality stands as a wide-ranging piece of innovative detailed research, full of new data and methods that were to be integrated into established geological procedure. Indeed, with the discovery of radioactivity, physicists and uniformitarian geology could final begin agreeing on the age of the earth, and Haughton's work on physical geology came into its own, but others took the credit for much of his laborious work.

The year 1859 marked a watershed for both science and Haughton. Coincidentally, the Carlowman enrolled at the advanced age of 38 as an undergraduate in the Medical School in Trinity, in the same year that Charles Darwin published *On the Origin of Species*. It was an incredible turn of events for an established mathematician and geologist, acclaimed for his work in both fields, and indeed already a Fellow of the Royal Society, to undertake the drudgery of attending a course of academic study. He continued to hold the chair of geology throughout his time as a medical undergraduate, and indeed there seems to have been very little sign of any let up in his prodigious publication record from this additional study load, as during this time he published 12 geological papers.

Clearly, at a personal level, he required a professional knowledge of comparative anatomy for his own developing research interests. Possibly the readiest way of acquiring this knowledge was through the study of human anatomy in the Medical School. There was, however, perhaps a second reason for his decision to register as an undergraduate. The university was seriously concerned about the standards of teaching in the Medical School. In particular, the Board was concerned about the well-reported slackness in directing dissecting classes and other practical work by the medical professors who, the Board believed, were more concerned with developing their own private practices than with teaching.

Haughton's appearance in the undergraduate class was certainly a shot over the bows of the errant medical professors. As it transpired, Haughton provided a long-term solution for the university to this problem. As soon as he graduated, he was appointed, in 1863, Registrar of the School of Medicine, and was thus in the administrative position to sort out the problem, which he, of course, did in short order.

After he graduated in the Medical School, Haughton's research also abruptly switched to zoology, and particularly to anatomical studies. He was easily able to obtain specimens from the Dublin Zoological Gardens, with which he was deeply involved from 1864. It was his habit to spend hours each day in the dissecting room, studying the comparative anatomy of the muscular systems of vertebrates. Haughton's reputation is a lasting one, and his book, *Principles of Animal Mechanics*, which appeared in 1873, was based on a decade of painstaking animal dissections. The work was for a considerable time the standard text on the topic, being the most comprehensive study of the physiological function of animals. The work included research on ostrich, emu, cassowary, pheasant, alligator, crocodile, hedgehog, monkey, llama, sloth, leopard, jaguar, tiger and lion. Haughton's use of advanced physics in explaining the movement of limbs and muscle systems is one of the most impressive and certainly the most original aspect of the book. He conclusively demonstrated that the 'Maupertius' principle of least action' applied to animal physiology.

The driving force in this study was his determination to undermine the theory of evolution, by removing the possibility of change in all animal species. He attempted to demonstrate that the animal forms were perfectly adapted to their environment, and therefore they would not evolve. He was, however, too late to make any crucial impact on the key evolutionary debate between T H Huxley and Sir Richard Owen, which had occurred a decade earlier. (By the 1870s, the issue that was at the storm centre of the debate, at which Huxley always contrived to keep himself, concerned bacteriological rather than human genesis.)

Haughton eventually published 31 papers on animal mechanics. His usual *modus operandi* in this period was to publish in the *Proceeding of the Royal Irish Academy*, and then to communicate the most important results to the Royal Society of London as a series of Discourses. At the end of the book *Principle of Animal Mechanics* he points out six major conclusions and, in Point 5, he claims to have demonstrated conclusively the permanence of each species, by the very criteria laid down by Darwin himself. A complimentary copy of his book sent to Darwin, described by Haughton as the 'author of the unproved hypothesis', was acknowledged, but all Haughton's laboriously crafted arguments appear to have solicited was a short reply: "I grieve that our theoretical views about the organic world differ so widely". His work also appears to generally have been ignored by the other evolutionists.

Haughton's own reputation posthumously suffered seriously from this role as a determined opponent of the theory of evolution. It is the victors who write the 'official' history, and they effectively wrote Haughton as 'a crossover biologist' out of the story. The same fate did not befall his collaborator Kelvin, who was fiercely protected by the physics community. Only now, as we begin to reassess these debates, and the more objective scientific merit of the work of the combatants, can we see the impressive nature of Haughton as both a biophysicist and biomechanist.

He is remembered in this regard particularly, if somewhat unfortunately, for his proposal for humane execution in 1866. His formula relates the weight of the unfortunate victim to the drop that is required to dislocate the joints at the junction of the vertebral column, and so damage the medulla oblongata to cause swift and therefore the desired painless death. (Haughton drop is measured in feet and is the weight of the felon in ounces divided by 2240.)

Haughton was particularly important as a medical reformer. As soon as he was appointed Registrar of the Medical School, he began the reform of the School by arranging the appointment of William Stokes as Chairman of the Medical Committee, with himself as Secretary. In 1864 he became a Governor of St Patrick Duns Hospital, and opened the hospital for surgical and medical patients. In 1867, the School of Physic Amendment Act, the so-called Haughton Act, passed into law.

Having copperfastened his power in Irish medical administration, he now set about the reform of Irish medical training. The degree of Bachelor of Surgery became registerable in 1876. He resigned from the position of Registrar of the Medical School in 1879, but only to make possible his appointment as the Chairman of the new Medical Committee, and thereby to extend his administrative power. Trinity pioneered the GMC examinations and registration, and Haughton played a key role in both the developments in Dublin (where he was naturally immediately appointed Trinity's GMC Representative) and also in London.

He surprisingly only published eight papers on medical topics and a mere four others on biological topics. For this polymath this was a relatively small number of papers, and demonstrated that his priorities were in medical administration. His legacy was, however, enormous in this field, as he brought about modern medical training in Ireland, and also played a significant role in the general reform of medical education in Britain.

His final phase of research concerned principally meteorological and chemical researches, and began in 1878. As usual, when he entered a new field of study, there was a rush of publication. By 1884 he had produced some 11 papers on solar radiation and related topics. His most important contribution adapted Rossetti's law of cooling to produce a new mathematical model for the relation of terrestrial radiation to the incoming solar radiation. He considered the effects of both the sun's and earth's heat inputs to the system, and also the effects of variation in atmospheric conditions. These factors were assessed in the context of the debate over the geological time, and the cause of glacial epochs and glacial climates. During this phase he produced 19 papers on solar radiation.

The final chapter of his scientific odyssey was a series of imaginative papers on chemistry, based on a Newtonian gravitational planetary

Samuel Haughton (courtesy of the provost, fellows and scholars of Trinity College Dublin).

model of molecules. The papers concerned the number of possible measurable quantities of the system, including angular momentum, the periodic rate of rotation, the molecular weight of the compound, the heat of formation of the molecule, and finally the size of the molecule. The series of seven papers dealt with binary, tertiary, quaternary and quinary compounds and appeared between 1887 and 1896. He also produced two papers on Newland's and Mendeleev's periodic laws of atomic weights.

Haughton became a member of the Royal Dublin Society and the Royal Irish Academy in 1845. He subsequently served on the Academy Council for some 30 years at various times and became Vice-President of the Academy in 1856. He was President from 1886 until 1891, during the very troubled times of the political dispute within the Academy. He was elected a Fellow of the Royal Society very early in his career in 1858. He was President of the Geological Society of Dublin from 1859, and President of the Dublin University Biological Society from 1876. He received only three honorary degrees, a DCL from Oxford in 1868, an LLD from Cambridge in 1880, and an MD from Bologna in 1888. He served the Royal Zoological Society from 1860 as a Committee member, then from 1864 as Secretary and finally from 1885 as President, and one may be forced to conclude that this Society was his real professional love.

Haughton was a man of religious conviction, and deeply attached to the Church of Ireland, having drifted from the Society of Friends after his marriage. He took holy orders in 1844 when he was elected to his Fellowship in Trinity. Haughton's views to the 1850–1851 Royal Commission were that the Queen's Colleges were 'impotent though angry and spiteful rivals of their Dublin sister'.

He was, however, given a very sharp wake-up call with the rise of continued advance of nationalism, which in the person of Dr Walsh, Archbishop of Dublin, had attacked Trinity as 'that ancient citadel of ascendancy and exclusiveness'. The defence of Protestant privilege changed fundamentally for both Galbraith and Haughton in 1868, with the election of Gladstone, who campaigned for the Disestablishment of the Church of Ireland. Galbraith was converted from a conservative and chaplain of the Grand Lodge into a Home Ruler.

This party was founded by Galbraith and Isaac Butt, both Trinity men, at a meeting in the Bilton Hotel in Dublin in May 1870, and it was indeed Haughton's closest collaborator that coined the name Home Rule. Soon Haughton had joined this political nationalist movement. Trinity was thereafter polarized into unionist and nationalist camps, and to a great extent so were the Royal Dublin Society and the Royal Irish Academy. Haughton's political star in fact rose in the 'nationalist' Academy, the desire for independence growing as the century progressed. During his life, he published only two articles on University reform, and the same number of sermons; for him, his polemical articles on the history of the earth were of far greater importance.

To summarize, Haughton was a founder of physical anatomy; he was a pioneer of earth modelling, considering both the internal motion of the molten core, and the meteorological issues relating to solar radiation, and significantly he entered the field before Kelvin; he was the world authority on tidal theory, and the geochronology of tidal sedimentation; he was a pioneer of tidal friction theories; he made many notable microscopic and other geological discoveries; he was an significant anti-evolutionist polemicist; he was a great educational innovator in scientific subjects; he was a master of administrative reform in both medical education and practice; he was an active participant in both the contemporary national and scientific political movements; but above all he was a Dublin University man to the last, but one of great courage and conviction, who stood for both Home Rule and reform, when these issues were very unpopular with a vociferous majority in Trinity.

Further Reading

'The Rev. Samuel Haughton: A man of Great Erudition and a Determined Opponent of the Theory of Evolution', N D McMillan, *Carloviana*, Part I: (1980) 12–14 ; Part II: (1981) 10–13. This article studies the Carlow background.

Rev. Samuel Haughton and the Age of the Earth Controversy, N D McMillan, in: *Science in Ireland 1800–1930*, J R Nudds, N D McMillan, D L Weaire and S McKenna Lawlor (eds) (Trinity College, Dublin 1988), pp 151–162.

Obituary, *Proceedings of the Royal Society of London* **62** (1897–8) 28–37 (written by D J Cunningham, Professor of Astronomy in Trinity College Dublin).

Obituary, *Proceedings of the Royal Irish Academy, Series 3,* Vol. IV (Appendix), (May 1898) 283–287 (written by D Spearman).

'Samuel Haughton: A Victorian Polymath', W J E Jessop, *Hermathena* **116** (1973) 5–26. This deals with Haughton's family connections.

William Thomson (Lord Kelvin)

1824–1907

Mark McCartney

William Thomson was born in Belfast on 26 June 1824. He was educated at Glasgow and Cambridge Universities, and from 1846 until 1899 he was Professor of Natural Philosophy at Glasgow University. He was elected a Fellow of the Royal Society in 1851, knighted in 1866, and elevated to the peerage as Lord Kelvin in 1892. William Thomson was arguably the most important scientist of the Victorian age. He was a leading figure in the creation of the area of physics concerned with heat and energy, known as thermodynamics, and was intimately involved in the laying of the first transatlantic telegraph cable. He corresponded with a range of other important scientists such as Stokes*, Fitzgerald*, Maxwell, Helmholtz and Joule. He died at his home near Largs in Ayrshire, Scotland on 17 December 1907.

In today's Belfast, the Georgian terrace of College Square East where William Thomson was born is a clutter of shop fronts looking out over congested traffic and a crowded city skyline. But when William was born, only two houses of the terrace stood, and before them, apart from the school where his father taught, there was only, in William's sister's words, "the open plain with its blue encircling hills".

William's father, James Thomson, the largely self-taught son of a Ballynahinch farmer, had studied hard and advanced far. He was Professor of Mathematics at the Belfast Academical Institution (today a school but, in the early days of its existence, a school and college combined with the college being essentially a Scottish university on a small scale) and he wrote a number of successful textbooks on topics such as arithmetic, calculus and trigonometry, which brought him a comfortable income.

William's early life in Belfast appears to have been as idyllic as the blue encircling hills which surrounded his home. His family was a close and affectionate one, and there were many friends. Family summer holidays included stays in coastal towns and villages around Belfast Lough; Donaghadee, Bangor and Carrickfergus were all backdrops to happy childhood memories.

In May 1830, however, the family circle was ruptured when William's mother died, and shortly after this the family moved to Glasgow, where

William's father had been appointed Professor of Mathematics at Glasgow University.

In 1834 William and his brother James both matriculated at Glasgow. James was 12, and William 10. This is remarkable, but not quite as remarkable as it may at first appear, as the usual minimum age for matriculation at Glasgow University was 14. In mathematics and natural philosophy William regularly came first in the class, and James second.

William's prodigious abilities in mathematics had been evident from an early age, and the obvious place where he should study after finishing at Glasgow was Cambridge. However, it was feared that if he graduated from Glasgow, he might not be able to enrol as an undergraduate at Cambridge. Thus although in May 1839 both boys passed the BA examinations, and in May 1840 both passed the MA examinations, on each occasion only James graduated. Around this time William would occasionally designate himself BATAIAP (Bachelor of Arts to all intents and purposes).

October 1841 saw William, aged 17, enter St Peter's College, Cambridge (Peterhouse) as a pensioner, i.e. a student who paid his own way. The formal tutoring in mathematics in his first year at Cambridge was of a very low level in comparison with what William already knew. By the time he reached Cambridge he had already published a paper in the *Cambridge Mathematical Journal*, defending the work of Fourier against the erroneous criticisms of Philip Kelland, Professor of Mathematics at Edinburgh University, and during his time as an undergraduate he wrote a further 10 papers. Indeed, virtually as soon as William had crossed the threshold of St Peter's, he was being tipped as Senior Wrangler (i.e. to come first in the final mathematics examinations).

While he was at Cambridge, every member of his family regularly wrote to him. His father, who was footing all the bills, often wrote on the wise use of money and time! In yearly college maintenance fees alone, the cost was £230. William's letters to his father would regularly contain detailed lists of all expenditure and, if writing for extra money, he would sometimes include a mathematical theorem for possible use in the Glasgow University exams to soften his father up.

In an early letter to his father William outlined his proposed daily schedule: rise at 5 a.m. to light his fire; read until 8.15 a.m.; attend daily lecture; read until 1 p.m.; exercise until 4 p.m.; chapel until 7 p.m.; read until 8.30, bed at 9 p.m. As Thomson's modern biographers, Smith and Wise point out, it is doubtful whether William actually adhered strictly to this, but it does illustrate his lifelong desire to minimize the waste of time.

At Cambridge, William involved himself in many other activities besides study: rowing (he was an excellent oarsman), practising on his cornet, helping to establish the University Music Society; as well as walking, skating and swimming. Of all these activities it was William's boating which his father disapproved of the most, fearing that it would bring his son into loose company, which would 'ruin [William] for ever' with wine parties and time wasting.

William's final examinations, the Senate House examinations, began on New Year's Day 1845, and lasted until 7 January. There were 12 papers, with morning papers two and a half hours long, and afternoon papers three hours long. The final result depended on both the quantity and quality of the answers to the questions. The exams formed the toughest mathematical racecourse in the land, with the competitors trained like thoroughbreds to answer the many bookwork questions at top speed, and to use all possible short cuts to get the answers.

To universal surprise William came not first but second. There was family disappointment, but justice was done when, in the Smith's Prize examination at the end of January, William came first. The Smith's Prize papers were more suited to William's abilities, as they contained more problem solving questions, and less of the bookwork which characterized the Senate House papers. Even though William had been placed second in the Senate House examinations, the comments around Cambridge realized him to be by far the greater mind, with one of the examiners commenting to a colleague: "You and I are just about fit to mend [Thomson's] pens." These successes meant that William was elected a Fellow of Peterhouse in June 1845 at the age of 21.

While William was studying at Cambridge, events were taking place in Glasgow which would define his future career. At the start of the 1841–1842 academic year at Glasgow, William's first year at Cambridge, the Professor of Natural Philosophy, William Meikleham was 70, ill, and not expected to return to his teaching post. It was part of William's father's duties, in his role as Professor of Mathematics, to start to think about Meikleham's replacement. He wanted academic excellence of the Cambridge calibre, but also a candidate who could teach well, and who had sympathy with the broad, non-hierarchical Scottish education system.

Initially James was concerned to secure a fellowship for William at either Cambridge or Trinity College Dublin, but as time wore on into 1843, and still no obvious candidate for the chair appeared, James started to see that William, then only 18, could be in the running.

Once the idea settled itself in James Thomson's mind, he began a careful and at times surreptitious campaign to have his son appointed to the Chair.

William's mathematical ability was not in doubt, but he needed experimental experience if he was to be in the running. His father advised him to get all the experience he could at Cambridge. Thus William attended lecture courses on experimental natural philosophy twice, once in 1843 and again in 1844. In 1844 he also attended the lectures on practical astronomy and astronomical instruments. Part of the manoeuvring also included a trip to Paris after he graduated in 1845; in Paris he attended lectures on chemistry and physics at the Sorbonne, and, on the advice of Dr William Thomson, one of his father's colleagues, bought and studied French texts. He met eminent men like Cauchy and Biot, and he worked for a time in the laboratory of Regnault, who was Professor of Natural Philosophy at the Collège de France.

In the summer of 1845 a new scenario arose. A new set of colleges was to be opened in Ireland, and it was being rumoured that the Presidency of Queen's College Belfast might well be offered to William's father. James was tempted by the idea, and of course William could then succeed him at Glasgow as Professor of Mathematics. But in the end, due to political manoeuvring by the Peel government to keep both the Catholics and Presbyterians happy, another candidate was offered the job, and James was offered instead the Vice-Presidency (at half the pay of the President). He refused, and the post was instead filled by Thomas Andrews*, one of his old pupils from the Belfast Academical Institution.

Rather conveniently, Professor Meikleham expired on 6 May 1846, just after William had finished his course at Cambridge. His illness had been in the background of William's thoughts for some time. In a letter to his father in 1844 he wrote, in a somewhat macabre passage: "Sorry to hear about Dr Meikleham's precarious state. I have now got so near to the end of my Cambridge course, that even on my own account I should be very sorry not to get completing it. For the project we have, it is certainly much to be wished that he should live till after the commencement of next session."

When Meikleham died, James wrote to William: "The enclosed notice [of Meikleham's death] must put you into active and energetic motion without delay."

From the moment Meikleham died, William and his father's covert manoeuvrings turned to overt action. A long list of testimonials were arranged, from the Master and Fellows of St Peter's, the Cambridge examiners, Augustus de Morgan, Arthur Cayley, Sir William Rowan Hamilton*, George Boole, George Gabriel Stokes, Victor Regnault and others. James wanted maximum impact; he arranged for testimonials to be printed and had a preference towards them being gilt edged. The printing would be done in Glasgow, and James would do the checking.

All was wreathed with success when, on 11 September 1846, William was unanimously elected, at the age of 22, to the Chair of Natural Philosophy at Glasgow. He held the post until 1899, and was not tempted away even by the Cavendish chair at Cambridge, which was offered to him three times in the 1870s and 1880s.

The first lecture of Professor William Thomson, given on 4 November 1846 was, he wrote to Stokes, a failure. It was all written down, and he read it too fast. Over the years students were to become devoted to their famous professor, but his quicksilver mind which saw connections and analogies across the discipline, and couldn't resist bringing them into lectures *ex tempore*, left a number of them floundering.

However, his enthusiasm, drive and fecundity transformed the natural philosophy course at Glasgow from a broad brush survey, to a thorough coverage of selected topics, with emphasis on modern developments. In his first five years he swept away much of the old equipment, spending £550 on new apparatus. He involved good students in experimental research, and at one point even extended his kingdom by stealth, taking over vacant college rooms, and only then asking the faculty if it was willing to ratify his actions retrospectively.

At this point it is perhaps appropriate to note the strength of the relationship between George Gabriel Stokes* and William Thomson. They met at Cambridge, and remained firm friends for the rest of their lives, exchanging over 650 letters. Much of this correspondence dealt with their researches in mathematics and physics. Their minds complemented each other, and in some cases their thoughts melded so that neither knew (nor cared) who had come up with an idea first. Perhaps the most famous example of this is the result known to all students of physics and mathematics as Stokes' theorem. Since it makes its first ever appearance in a letter *from* William *to* Stokes, it should really be called Thomson's theorem.

In 1847 William came across the Mancunian James Prescott Joule at the meeting of the British Association in Oxford. Since 1843, Joule had been claiming at the British Association meetings that heat was not, as was supposed at the time, a substance, known as caloric, which moved between materials, but was a state of vibration of its atomic constituents. At previous meetings he had suggested a zero of temperature at −284°C, based on the decrease in volume of a gas with decrease in temperature. He had demonstrated the equivalence of work and heat, by use of experiments to determine the equivalent amount of mechanical work required to raise one pound of water through 1°F; and he had suggested that the temperature of water at the bottom of a waterfall should be greater than at the top.

Joule's talks at the British Association were, however, met with the silence of apathy and incredulity. At least that was the case, until the day in 1847 when William Thomson was in the audience. William asked questions from the floor, and provoked a debate. William became interested, but assumed that Joule must be wrong. (After the Oxford meeting, he wrote to his brother: "I enclose Joule's papers, which will astonish you. I have only had time to glance through them as yet. I think at present some great flaws must be found.'') But Joule was not wrong, and William, through careful thought, came to agree with him. Along the way, he connected Joule's work with that of Carnot on heat engines; he devised a more fundamental way to define the absolute zero of temperature, independent of any particular material substance (and it is for this reason that the fundamental unit of temperature is called the kelvin); he saw the idea of conservation of energy as a great unifying principle in science; and he introduced the ideas of statical and dynamical energy, or what we would call potential and kinetic energy.

It is difficult to disentangle William's work on heat and the conservation of energy from that of other scientists of the time, including Joule, Clausius, Rankine, Liebig and Helmholtz. All of them can take some of the credit for the first and second laws of thermodynamics, and these ideas are so important to modern science that each contributor should be held in high regard.

In June 1851, just before his 27th birthday, William's scientific achievements were recognized by his election as a Fellow of the Royal Society. In September 1852, aged 28, William married his second cousin Margaret Crum. William proposed to Margaret on the rebound from three rejected proposals of marriage to one Sabrina Smith. His last proposal of marriage to Sabrina was in April 1852, and he proposed to Margaret in July of the same year!

William was extremely highly regarded as a pure scientist, but perhaps became even more famous as a result of his applications of science. December 1856 saw the formation of the Atlantic Telegraph Company, with William on the board of directors. Public interest in the project to lay a telegraph cable from Britain to America was great and William over the period 1854–1858 gradually became a central public figure.

The first attempt at laying a cable was in 1857 was a failure, and it wasn't until the fifth attempt, in 1866, that a cable was laid successfully. The *Times* called the success ''the most wonderful achievement of this victorious century''. The link was between Newfoundland in Canada, and Valentia in Ireland. William had spent months at sea, and had become intimately and enthusiastically involved with the practicalities of the project. As a result of his work he patented a number of devices, and also formed a firm of consulting engineers. Along with others involved in the project, William was knighted on 10 November 1866.

William's views on the thermal history of the earth also became extremely well known. His interest in this topic began in 1844, while he was still a Cambridge undergraduate. It was a topic he returned to repeatedly, and which led him into conflict with other scientists, such as John Tyndall*, T H Huxley and Charles Darwin. The tone of the debate was not advanced by Darwin describing William as an 'odious spectre', nor by Huxley promoting evolutionary theory as an alternative to religious

Thomson in 1859 at the age of 35. Photograph by the Rev. David King, Thomson's brother-in-law. Thomson is reading a letter from Fleeming Jenkin. From Agnes Gardner King 1925 *Kelvin the Man* (London: Hodder and Stoughton) frontispiece.

belief with evangelical fervour. William was a Christian, but he was not concerned with defending a literal interpretation of the Genesis narratives; indeed he was happy to speculate that life came to earth via a meteor. He was, however, concerned with defending and promoting good science. He believed geology and evolutionary biology were weak subjects when up against the rigours of mathematically based natural philosophy. Indeed many physicists didn't even believe that geology and biology were sciences at all. To evaluate the age of the earth, he used the methods of his beloved Fourier to calculate how long it had taken the earth to cool from a molten state to its current temperature.

William had read Fourier's book, *The Analytical Theory of Heat*, at the age of 16, and had fallen in love with it. In some ways it set the agenda for many aspects of his life's research. The mathematical description of heat flow linked his work on thermodynamics, the cooling of the earth, and even the flow of electrical signals through telegraph wires; in each case William attempted to cast the problem in terms which could be solved using Fourier's methods. In the case of electrical signals through telegraph wires, however, William's love of Fourier initially lead him astray. The equation he initially proposed had the beautiful property that it was a direct analogue to Fourier's equation for heat diffusion. It was also wrong, however, as it completely ignored self-inductance.

To the chagrin of the biologists and geologists, William's calculations for the age of the earth didn't allow enough time for evolution to occur. The discrepancy between the theories was not resolved until the beginning of the twentieth century, when Ernest Rutherford realized that radioactivity provides the earth with an internal heating mechanism, which opposes, and thus slows down, the cooling, and so lengthens the time scale sufficiently. Given that radioactivity was not discovered until William was in his seventies, he can be forgiven for not using it in research carried out in his twenties and thirties!

On 17 June 1870, William's wife Margaret died; she had been ill for virtually the whole of her married life. Three months after her death William bought a 17-year-old 126-ton sailing boat, *Lalla Rookh*. This provided a diversion from his bereavement, and since he spent much time aboard her pursuing his researches, it also gave a change of scene from familiar places which were perhaps haunted by the memory of his wife.

Time spent on *Lalla Rookh* opened up William's interest in navigation, and he designed and patented a new compass, which was more stable than existing ones, and compensated for the effect of the iron hull of modern ships. Initially the Admiralty were sceptical, with one committee claiming it to be 'too flimsy, sure to be fragile'. William's response was to throw his compass across the committee room; it remained intact. A probably apocryphal addition to this story says that William then threw the admiralty standard issue compass across the room. It, so the story goes, didn't survive the impact. The Navy were finally convinced of the soundness of the new compass, and by 1888 it was adopted as standard on all Admiralty ships. Apart from the compass, William also invented a mechanical tide predictor, and developed a new sounding machine which allowed depths to be taken quickly and, more importantly, without having to stop the ship.

William's continuing work on laying submarine telegraph cables took him to Madeira in 1873, but the ship he was on had to remain off the coast of the island for some 16 days, due to a fault on one of the cables. During this time William was entertained by the island's wealthiest landowner, Mr Blandy. In May of 1874 William sailed back to Madeira on the *Lalla Rookh*, and proposed to Mr Blandy's second daughter, Fanny. She accepted and they were married on William's fiftieth birthday.

Soon after his second marriage, William and his new wife built a new home, Netherhall, at Largs near Glasgow. It was largely designed by William himself, and was in the Scottish Baronial style, complete with peacocks for the grounds supplied by James Clerk Maxwell. True

Thomson in 1897 at the age of 73. Photograph by Annan, Glasgow. From Silvanus P Thompson 1910 *The Life of William Thomson, Baron Kelvin of Largs* 2 vols (London: Macmillan) vol. I frontispiece.

to William's love of inventions and progress, Netherhall was one of the first houses to be fitted with electric light.

In 1884 William, now aged 60, and Fanny went to North America. The meetings of the British Association were being held in Montreal, and he was then to give a series of 20 lectures at Johns Hopkins University in Baltimore. His discursive style of lecturing, with frequent digressions onto other issues, a trait which some of his students in Glasgow found positively confusing, was considered 'delightful' by the North Americans. William's Baltimore lectures were interactive and spontaneous; he did not have them previously prepared. Indeed Lord Rayleigh, who along with men like Michelson and Morley, was in the audience, noticed that many of the morning lectures were based on questions that had been raised by the group over breakfast. William spoke on subjects such as the wave nature of light, the existence of the ether (even going so far as to calculate its mass/cubic km) and the nature of the atom.

In 1892, aged 68, William was raised to the peerage, becoming Baron Kelvin of Largs. Four years later, in 1896, William's 50th jubilee as Professor at Glasgow was celebrated in grand style. The affair lasted three days, with greetings from the Queen, the Prince of Wales and a wide range of academics, scientific societies and engineers.

William resigned his chair in 1899 after 53 years as Professor of Natural Philosophy at Glasgow, and true to his inquisitive and enthusiastic character, promptly enrolled himself as a research student, thus making himself one of the youngest and oldest students in the university's history.

William died at his home on 17 December 1907 and was buried in Westminster Abbey near Isaac Newton.

Further Reading

Lord Kelvin's Early Home, E T King (Macmillan, London, 1909).

Life of Lord Kelvin, S P Thompson (Macmillan, London, 1910).

Lord Kelvin: The Dynamic Victorian, H I Sharlin (Pennsylvania State University Press, 1979).

Energy and Empire: A Biographical Study of Lord Kelvin, C Smith and M N Wise (Cambridge University Press, 1989).

George Johnstone Stoney

1826–1911

James G O'Hara

Johnstone Stoney was born on 15 February 1826 at Oakley Park, Clareen, near Birr in King's County (now Offaly). The Stoneys, a long-established Anglo-Irish family, lost their property during the famine years and moved to Dublin to allow the children to pursue professional careers. George Johnstone graduated in 1848 from Trinity College, excelling in mathematics and physics, before spending two and a half years as astronomical assistant to Lord Rosse* at Parsonstown (Birr Castle) Observatory. While there he studied for the highly competitive Trinity College Fellowship, which would have opened up a scientific and academic career for him in Dublin but, in the contest of 1852, he only attained second place, winning the Madden prize. Through the influence of William Parsons, Lord Rosse, he was appointed Professor of Natural Philosophy at Queen's College Galway, where he remained for five years until he became Secretary to the Queen's University, based in Dublin Castle. Stoney retained this post until the Queen's University was dissolved in 1882 and replaced by the Royal University, which was a purely examining body and the predecessor of the National University of Ireland, and the Queen's University of Belfast.

In the capacity of Secretary to the Queen's University and Superintendent of Civil Service Examinations in Ireland, Stoney devoted the greater part of his professional career to university policy and administration. At the same time, he was able to continue scientific and scholarly work through his close association with the Royal Dublin Society (hereafter abbreviated as RDS). From 1871 he was Honorary Secretary, and from 1881 Vice President, of the Society; much of his own research work was communicated to the Society and published in its scientific journals, and in 1899 he was honoured with the Society's first Boyle medal.

In 1893 at the age of 67, Stoney left Ireland and lived in retirement in London, intending to provide his children with better educational and career opportunities. His life there was largely devoted to the completion and publication of work begun in Ireland. Stoney died in London at the age of 85 on 5 July 1911. His body was cremated, and his ashes buried at Dundrum, Dublin.

In physics, Stoney's first studies were concerned with the kinetic theory of gases. When, in about 1860, he turned his attention to this theory, some important results had recently been published. In 1857, Rudolf Clausius in Germany had given an expression which allowed the determination of a mean velocity of gas molecules under ambient conditions. Then in 1859 he had found an expression for the mean free path of a molecule in terms of the mean molecular separation; where all the molecules of a hypothetical gas in a container move with the same velocity, the mean free path was found to be about 60 times the mean separation.

In 1859 the Scottish physicist James Clerk Maxwell had published a theory based on a distribution of molecular velocities, which he then applied to the results of experiments on gaseous viscosity and diffusion, in order to obtain real numerical values for the mean free path of molecules. In his first significant paper on gas theory, in 1868, Stoney combined these results to obtain a further important quantity. From an average of three values for the mean free path published by Maxwell, Stoney found the mean separation of gas molecules, at atmospheric pressure and room temperature, to be of the order of 10^{-6} mm, being about one sixtieth of the mean free path according to Clausius' result. He could therefore deduce that the number of molecules in a cubic mm of gas at atmospheric pressure and room temperature was of the order of 10^{18}, a quantity closely related to the still undetermined Avogadro's constant.

The main intention of this 1868 paper, entitled *On the Internal Motions of Gases Compared with the Motions of the Waves of Light*, was to connect data about molecular magnitudes obtained from the kinetic theory, with the wavelengths and frequencies of light. The kernel of Stoney's theory was his hypothesis that light waves arise from periodic motions within the atoms or molecules. He suggested each molecule was a complex system, having two distinct forms of motion, namely translatory and internal motions, and, elaborating on earlier considerations of Clausius, he maintained that the 'vis viva', expressed mathematically by the formula mv^2, of these internal motions constitutes a definite part of the total heat of the gas. The internal motions were complex but regular orbital motions which affect the surrounding ether, producing light of definite wavelengths. He concluded that between 50 000 and 100 000 orbital motions would be executed on average between molecular encounters. The internal motions could be resolved into a number of simple harmonic vibrations of definite periods.

In studies of electrochemistry, Stoney developed the concept of an atom or quantum of electricity. The origins of this concept in electrolytic theory can be traced back to the investigations of Michael Faraday in 1833–1834, and James Clerk Maxwell in the 1860s. However, as regards the actual existence of an atom or molecule of electricity, both Faraday and Maxwell were extremely sceptical.

Following the publication of Maxwell's celebrated *Treatise on Electricity and Magnetism* in 1873, Stoney gave his interpretation of electrolysis in a paper presented, in August 1874, at the Belfast meeting of the British Association for the Advancement of Science (hereafter abbreviated as BA). Only the title of the paper, *On the Physical Units of Nature*, was published in the Report of that meeting; seven years later, on 16 February 1881, Stoney read a paper with the same title before the RDS and it was published soon after in the *Philosophical Magazine*. Here he discussed three systems of units, namely the BA ohm series, a metric metre–gramme–second (MGS) system, and a system of natural units that he had devised himself.

The first system had been developed by a committee of the Association first appointed in 1862 to establish a unit of electrical resistance. By 1874 the work of this committee, whose members included Maxwell, William Thomson* (later Lord Kelvin) and Stoney, had been extended, and a practical system of units for electrical notation developed. Based on arbitrary units of length, mass and time (10^7 m, 10^{-11} g, and 1 s) a system of electrical units had been proposed which included the 'ampère' (unit of quantity), the 'ohm' (resistance), the 'weber' (current), the 'volt' (emf) and 'farad' (capacity). Despite some similarities, this is clearly very different from the system we use today.

Parallel to this system Stoney discussed a second arbitrary MGS system. In this connection he introduced a special nomenclature designating the fundamental and derived units as 'lengthine', 'massine', 'timine', 'forcine', 'velocitine', etc. The unit of quantity of electricity or charge was thus the 'electrine' and was equivalent to 100 ampères in the BA ohm series.

The third system of units proposed by Stoney was a natural system whose fundamental units were not arbitrary units, but rather physical constants. The three units chosen were derived from electromagnetism, gravitation and electricity, and designated V_1, G_1 and E_1, respectively. V_1 is the ratio connecting electrostatic and electromagnetic quantities in a medium of inductive capacity unity; it has the dimensions of a velocity and is numerically equal to the maximum velocity of electromagnetic radiation. G_1 is the coefficient of universal gravitation, and E_1 is the fundamental unit of quantity of electricity. Stoney later proposed the names 'Maxwell' and 'Newton' for V_1 and G_1, respectively, and 'electron' for the third unit E_1.

As regards this third fundamental unit he wrote in 1881: "And, finally, Nature presents us, in the phenomenon of electrolysis, with a single definite quantity of electricity which is independent of the particular bodies acted on. To make this clear I shall express 'Faraday's Law' in the following terms, which, as I shall show will give it precision, viz.: for each chemical bond which is ruptured within an electrolyte a certain quantity of electricity traverses the electrolyte which is the same in all cases. This definite quantity of electricity I shall call E_1. If we make this our unit quantity of electricity we shall probably have made a very important step in our study of molecular phenomena.... Now the whole of the quantitative facts of electrolysis may be summed up in the statement that A DEFINITE QUANTITY OF ELECTRICITY TRAVERSES THE SOLUTION FOR EACH CHEMICAL BOND THAT IS SEPARATED''.

In his 1881 paper Stoney determined for the first time the mass and charge of the electrolytic hydrogen ion in the following way. The mass of a litre of hydrogen was about 0.1 g, and, from his 1868 result, the number of molecules in this quantity would be about 10^{24}; the mass of the hydrogen molecule must therefore be about 10^{-25} g. The mass of the chemical atom would of course only be half that of the molecule, but he decided to ignore this considering the approximate nature of the data.

From experiments on the electrolysis of water, the BA committee had established that every 'ampère' of electricity effects the decomposition of 92×10^{-6} g of water, yielding about 10^{-5} g of hydrogen. His 'electrine', i.e. 100 'ampère', would thus liberate about 1 mg of hydrogen. This quantity would contain about 10^{22} atoms or molecules, which represents the number of bonds broken by the passage of an 'electrine'. The quantity of electricity (E_1) corresponding to each bond ruptured is therefore found by dividing the value of the 'electrine' (100 'ampère') by 10^{22} which gives a value of 10^{-20} 'ampère'. This value is about one-sixteenth of the correct value for the charge of the electron, i.e. 1.6×10^{-19} coulomb or 4.8×10^{-10} CGS esu.

Seven weeks after Stoney read his paper at the RDS in 1881, and a month before it was published in the *Philosophical Magazine,* a much more famous contemporary, Hermann von Helmholtz, also advocated the acceptance of the atom of electricity. Helmholtz, the patriarch of German science, had been invited to deliver a Memorial lecture, commemorating the commencement of Faraday's investigations half a century earlier, to the Fellows of the Chemical Society at the Royal Institution, London, on 5 April 1881. On this occasion Helmholtz gave a corpuscular interpretation of Faraday's laws of electrolysis, and he asserted that ''electricity... positive as well as negative, is divided into definite portions, which behave like atoms of electricity''. Helmholtz's lecture attracted much attention, and created a much more favourable climate for the reception of such ideas.

Spectroscopy became an exciting research field in the middle decades of the nineteenth century, and in this area too Stoney made significant contributions. It was in the context of the quest for exact numerical relationships, that might lead to mathematical laws for the distribution of spectral lines, and for a corresponding atomic model, that his work was mainly carried out. Already in his 1868 paper on kinetic theory he had suggested that the periodic motions within atoms or molecules could be the source of spectral lines. However, it was in a paper of 1870, *On the Cause of the Interrupted Spectra of Gases*, that he first began to grapple with the fundamental problems of how to explain the structure of line spectra. Spectral lines were generally thought to be of atomic or molecular origin, but, if each line corresponded to a specific movement of the emitting molecule, how could this vibrate in so many different ways?

Stoney adopted an analogy with acoustics, namely of a fundamental note and the harmonics produced by a vibrating string. He considered that a complex undulation would be impressed on the ether by one of the periodic motions of a gaseous atom. In mathematical terms this waveform was analysed, and found to consist of a simple harmonic motion of a certain fundamental period, together with a series of harmonics. As long as the undulation travels through a vacuum it retains its complex form, but on entering a dispersing medium the constituents separate, each travelling at a speed proportional to its period. If the dispersing medium were to be shaped as a prism, the constituents would emerge in different directions, each appearing as a line in the spectrum. Thus a single molecular vibration could give rise to a series of lines in a spectrum, and the existence of a number of fundamental motions in a molecule would account for the scattered appearance of lines observed in spectra.

Stoney applied this theory to the visible hydrogen spectrum, which had its four known lines, corresponding to four lines observed in the solar spectrum. He was able to attribute three of the four to a single internal molecular motion. He found them to be the twentieth, twenty-seventh and thirty-second harmonics, respectively, of a certain fundamental vibration. The fourth line did not fit into the series of harmonics, nor were any further harmonics to be found. He believed, however, that these would in time be discovered. He also used the theory of harmonic series to explain the appearance of fluted band spectra, suggesting that these patterns were made up of lines, or harmonics, ruled in very close succession. Stoney's paper, first presented to the BA in 1870, was published in the following year in the *Philosophical Magazine.*

Shortly after, he published a second part of this inquiry jointly with James Emerson Reynolds, the analyst of the RDS and keeper of its Mineral Department. They carried out a laboratory investigation of the absorption spectrum of the vapour of chromyl chloride (CrO_2Cl_2), whose lines they thought might be accounted for by a single molecular motion. In testing the theory of simple harmonic ratios Stoney and Reynolds used a scale of inverse wavelengths in preference to ordinary wavelengths. This proved to be a significant innovation in spectroscopy; harmonics, where they existed, would be equally spaced on such a scale, and the interval between successive harmonics would equal the value of the 'fundamental'. With a spectroscope made by the Dublin instrument makers Thomas and Howard Grubb* they measured the angular deviations of 31 of the 106 lines observed in the spectrum of chromyl chloride, deduced their inverse wavelengths and concluded that they represented a harmonic series.

Stoney's theory of simple harmonic ratios and harmonic series dominated research efforts for the next decade. At the 1871 meeting of the BA, Stoney outlined the advantages of an inverse scale, and he presented a table of inverse wavelengths or wave numbers for the principal lines of the solar spectrum. A committee was set up to produce a catalogue of spectral rays on a wave number scale, and at the Dublin meeting of the Association in 1878 the catalogue was presented. In the meantime other physicists discovered and published harmonic series for other elements similar to Stoney's hydrogen series.

The discovery of hitherto unknown lines in the ultra-violet region of the hydrogen spectrum towards the end of the decade presented a new challenge. Hermann Wilhelm Vogel in Berlin discovered five new hydrogen lines in 1879, and in the following year William Huggins in England discovered 12 strong lines in the spectra of a number of white stars, three of which coincided with known hydrogen lines of the solar spectrum. Huggins communicated his results to Stoney, and included the latter's reply when he published his results. The central issue for Stoney was whether or not these lines could be related by a single law, that is whether or not the points lay exactly on or just close to a specific curve. He therefore plotted all 14 lines against their wave frequencies, or the reciprocals of the wavelengths, and he concluded incorrectly that the points were close to, but not exactly on, a definite curve, and therefore that Huggins' lines were to be regarded, not as consecutive members of a single series, but rather as members of one or more series with positions near the curve.

Stoney thus failed to deduce the law for the hydrogen series and the first spectral series formula. His belief in the law of harmonic ratios seems to have predisposed him to the view that an exact mathematical law did not exist. In the end it was Johann Jacob Balmer, a 60-year-old Swiss schoolmaster, who solved this mathematical puzzle. Stoney's simple numerical relationships had convinced Balmer that a simple law must exist, and he eventually succeeded in relating all four lines in the visible hydrogen spectrum by the four proper fractions to a key-note or 'fundamental number'. Even after the publication of Balmer's Law, Stoney continued to insist on the validity of his relationship of 1870–1871 for the three hydrogen lines which he regarded as a particular case of this law.

Following Heinrich Hertz's verification, in 1887/1888, of Maxwell's prediction that light is a form of electromagnetic radiation, physicists began to think in terms of an electromagnetic explanation of optical and spectral phenomena. At this juncture, the atom or quantum of electricity, as conceived by Stoney, was to have a new lease of life. The most elaborate and most important formulation of Stoney's thinking is found in a paper, *On the Cause of Double Lines and of Equidistant Satellites in the Spectra of Gases*, which he read at the RDS on 26 March and 22 May 1891. Stoney once again argued that spectral phenomena arose from undulations in the luminiferous ether created by periodic vibrations occurring within molecules. He held both gaseous molecules and chemical atoms to be elaborate systems, within which highly complex but periodic changes were constantly taking place.

He attempted to account for line spectra using a kinematical model, and treating the molecule as a planetary system. A point, acting on the ether, was considered to move along an orbit in a molecule following an intermolecular encounter. The actual course of the point is found by first establishing the dominant orbit, and then subjecting it to perturbations. In order to give an electromagnetic interpretation, he resorted to his earlier conception of a quantum of electric charge, referred to now as an 'electron'. This was the first time the term was used in modern physics. According to Stoney, there are at least two (and possibly several) such 'electrons' in each atom; furthermore an electron cannot be removed from an atom, except where the chemical bond is broken in electrolysis.

Stoney's 'electron' was a unit of charge, and not a subatomic particle; neither the sign of its charge nor its mass play a role. The motions of the electrons in perturbed elliptical orbits create the electromagnetic undulations in the ether, which the spectroscope resolves as spectral lines. The actual orbit of an electron might be complex, but, in mathematical treatment, it could be resolved into elliptical 'partials'. An apsidal shifting of the partial in its plane would give rise to a double line in the spectrum, and the observation of series of doublets could accordingly provide information about the electronic orbit. Stoney attempted to apply his model to account for the line spectra of hydrogen and the alkali metals, each of which was known to consist of three series of double lines. He also considered a number of other perturbations and the corresponding effects that would result in the spectrum.

In 1876 Stoney attempted an explanation of William Crookes' radiometer which, in its original form, consisted of an arm with pith balls at its ends and mounted on a pivot in a vacuum tube. He suggested that, in a very rarefied state when the mean free paths of molecules would be inordinately large, the residual gas causes the movement of the vanes. Adopting Stoney's idea, Crookes himself proposed that a 'fourth state of matter' is attained when the mean free path is of the order of the dimensions of the vessel. In 1882 Crookes then applied this concept of the fourth state, or radiant matter, to electrical effects in rarefied gases in general; this in turn initiated the so-called cathode-ray controversy—the debate

George Johnstone Stoney (courtesy of the Royal Society, London).

about the particulate or wave nature of cathode rays—that was only resolved by the experimental discovery of the electron in 1897.

Certain of Stoney's contributions in physical optics, solar and atomic physics, some of which he considered to be his finest achievements, were not well received by contemporaries or appreciated by posterity, being seen as unconventional, off-beat contributions by an amateur scientist. In cosmic physics, he investigated a planet's power of retaining its gaseous envelope, and his work suggested that hydrogen and helium, the existence of which had been inferred from the solar spectral observations, must eventually escape from the earth's atmosphere. Starting with a substantial paper on the physical constitution of the sun and the stars, he published a series of theoretical contributions between 1868 and 1897 on the conditions limiting planetary and lunar atmospheres.

In 1888 he undertook an investigation of the numerical relations of atomic weights; the Royal Society published an outline, but full publication was rejected, a matter which greatly grieved him. His main idea was that if a succession of spheres be taken, whose volumes are proportional to atomic weights, and if these atomic spheres be plotted as ordinates against a series of integers as abscissae, a logarithmic curve is produced. His theory predicted the existence of missing elements corresponding to the family of inert gases later discovered and isolated, mainly by William Ramsey between 1894 and 1900. Stoney upheld his priority claim in a letter to the *Philosophical Magazine* in October 1902, and in a memorandum published by Lord Rayleigh in the *Proceedings of the Royal Society* in July 1911, the month of Stoney's death.

Stoney's place in the history of science is first and foremost connected with the discovery of the electron. The history of this discovery can be treated in two periods: the speculative and empirical investigation of the concept of an atom of electricity, which extends over a period of about 60 years from Faraday's pioneering investigations in 1833/1834, and the following period of three to five years when the crucial experimental investigations, culminating in J J Thomson's studies of cathode rays, were undertaken. In the first period Stoney stands out as a major contributor; in the second period he was superseded by a younger generation of professional physicists, among them Irishmen like Stoney's nephew George Francis FitzGerald*, Thomas Preston and Joseph Larmor*. The work of each of these physicists was coloured in some way through their association with and the influence of Johnstone Stoney, as he was he was affectionately called by contemporaries.

Further Reading

Dictionary of National Biography and the forthcoming *New Dictionary of National Biography*.

'George Johnstone Stoney FRS and the Concept of the Electron', J G O'Hara, *Notes and Records of the Royal Society* **29**(2) (1975) 265–276.

'George Johnstone Stoney and the Conceptual Discovery of the Electron', J G O'Hara in: Occasional Papers in Irish Science and Technology 8: *Stoney and the Electron*, Royal Dublin Society, 1993, pp 5-28.

'George Johnstone Stoney, 1826–1911', J Joly, *Proceedings of the Royal Society* **86A** (1912) xx–xxxv.

George Francis FitzGerald

1851–1901

J J O'Connor and E F Robertson

George FitzGerald was a brilliant mathematical physicist, who today is known by most scientists as one of the proposers of the FitzGerald–Lorentz contraction in the theory of relativity. However, this suggestion by FitzGerald, as we shall see below, was not in the area in which he undertook most of his research, and he would certainly not have rated this his greatest contribution.

George FitzGerald's parents were William FitzGerald and Anne Frances Stoney. His father William was a minister in the Church of Ireland, and rector of St Ann's, Dublin, at the time of George's birth. William, although having no scientific interests himself, was an intellectual who went on to become Bishop of Cork, and later Bishop of Killaloe. It seems that George's later interest in metaphysics came from his father's side of the family. George's mother was the daughter of George Stoney from Birr in King's County, and she was also from an intellectual family. George Johnstone Stoney*, who was Anne's brother, was elected a Fellow of the Royal Society of London and George FitzGerald's liking for mathematics and physics seems to have come mainly from his mother's side of the family.

William and Anne had three sons, George being the middle of the three. Maurice FitzGerald, one of George's two brothers, also went on to achieve academic success in the sciences, becoming Professor of Engineering at Queen's College Belfast. George's schooling was at home where, together with his brothers and sisters, he was tutored by M A Boole, who was the sister of George Boole, the eminent mathematician. It is doubtful whether Miss Boole realized what enormous potential her pupil George had, for although he showed himself to be an excellent student of arithmetic and algebra, he was no better than an average pupil at languages, and had rather a poor verbal memory. However, when the tutoring progressed to a study of Euclid's *Elements*, George showed himself very able indeed, and he also exhibited a great inventiveness for mechanical constructions, having great dexterity. He was also an athletic boy but he had no great liking for games.

Miss Boole prepared her pupils very well for their university studies. She noticed one remarkable talent in her pupil George; that was his skill as

George Francis FitzGerald (courtesy of AIP Emilio Segrè Visual Archives, Physics Today Collection).

an observer. Many years later FitzGerald, clearly thinking of his own youth, wrote: "The cultivation and training of the practical ability to do things and to learn from observation, experiment and measurement, is a part of education which the clergyman and the lawyer can maybe neglect, because they have to deal with emotions and words, but which the doctor and the engineer can only neglect at their own peril and that of those who employ them. These habits should be carefully cultivated from the earliest years while a child's character is being developed. As the twig is bent so the tree inclines."

FitzGerald certainly showed that he had acquired the ability to learn from observation, experiment and measurement. He entered Trinity College Dublin at the young age of 16 to study his two best subjects, which were mathematics and experimental science, and he was soon putting the training he had received at home to good use. At Trinity College, FitzGerald 'attained all the distinctions that lay in his path with an ease, and wore them with a grace, that endeared him to his rivals and contemporaries'.

It was not an undergraduate career devoted entirely to study, however, for FitzGerald played a full part in literary clubs and social clubs. He also continued his athletic interests, taking to gymnastics and to racquet sports. In 1871 he graduated as the best student in both mathematics and experimental science. He won a University Studentship and two First Senior Moderatorships in his chosen topics.

The aim of FitzGerald was now to win a Trinity College Fellowship, but at this time these were few and far between. He was to spend six years studying before he obtained the Fellowship he wanted, but during these years he laid the foundation of his research career. He studied the works of Lagrange, Laplace, Franz Neumann, and those of his own countrymen Hamilton* and MacCullagh*. In addition he absorbed the theories put forward by Cauchy and Green. Then, in 1873, a publication appeared which would play a major role in his future. This was *Electricity and Magnetism* by James Clerk Maxwell, which, for the first time, contained the four partial differential equations now known as Maxwell's equations. FitzGerald immediately saw Maxwell's work as providing the framework for further development and he began to work on pushing forward the theory.

It is worth noting that FitzGerald's reaction to Maxwell's fundamental paper was not that of most scientists. Very few seemed to see the theory as a starting point, rather most saw it only a means to produce Maxwell's own results. It is a tribute to FitzGerald's insight as a scientist that he saw clearly from the beginning the importance of *Electricity and Magnetism.* Maxwell's theory was, for many years, in the words of Heaviside, 'considerably underdeveloped and little understood', but a few others were to see it in the same light as FitzGerald, including Heaviside, Hertz and Lorentz. FitzGerald would exchange ideas over the following years with all three of these scientists.

During the six years he spent working for the Fellowship, FitzGerald also studied metaphysics, a topic which he had not formally studied as an undergraduate, and he was particularly attracted to Berkeley's philosophy. His liking for metaphysics, and his deep understanding of the topic, combined with his other great talents in his future career. He failed to win a Fellowship at his first attempt, but succeeded at his second in 1877, and became a tutor at Trinity College Dublin. At Trinity College he was attached to the Department of Experimental Physics, and soon he was the major influence on the teaching of the physical sciences in the College.

In 1881 John R Leslie, the Professor of Natural Philosophy, died, and FitzGerald succeeded him to the Erasmus Smith Chair of Natural and Experimental Philosophy. At the time of his appointment he gave up his duties as College tutor, a role in which he had been extremely successful, to concentrate on his duties as a professor. One of FitzGerald's long running battles at Trinity College Dublin was to increase the amount of teaching of experimental physics. He soon set up classes in an old chemical laboratory that he was able to obtain for his use, and he gathered around him colleagues who would help in the practical aspects of the subject. As is so often the case in universities, however, he was restricted in the progress he could make from a lack of funds.

FitzGerald attempting to fly in the College Park of Trinity College Dublin in the 1890s.

In a lecture which he gave to the Irish Industrial League in 1896 FitzGerald emphasized his lifelong belief in practical studies: "The fault of our present system is in supposing that learning to use words teaches us to use things. This is at its best. It really does not even teach children to use words, it only teaches them to learn words, to stuff their memories with phrases, to be a pack of parrots, to suffocate thought with indigestible verbiage. Take the case of experimenting. How can you teach children to make careful experiments with words? Yet it is great importance that they should be able to learn from experiments."

However, practical applications are built on theoretical foundations, and FitzGerald fully understood this. In his inaugural lecture as President of the Dublin Section of the Institution of Electrical Engineers, on 22 February 1900, he spoke of how electricity had been applied to the benefit of mankind during the nineteenth century. Behind a practical invention such as telegraphy there was a wealth of theoretical work: "Telegraphy owes a great deal to Euclid and other pure geometers, to the Greek and Arabian mathematicians who invented our scale of numeration and algebra, to Galileo and Newton who founded dynamics, to Newton and Leibniz who invented the calculus, to Volta who discovered the galvanic coil, to Oersted who discovered the magnetic actions of currents, to Ampère who found out the laws of their action, to Ohm who discovered the law of resistance of wires, to Wheatstone, to Faraday, to Lord Kelvin*, to Clerk Maxwell, to Hertz. Without the discoveries, inventions, and theories of these abstract scientific men telegraphy, as it now is, would be impossible."

We should also look at FitzGerald's idea of the purpose of a university, for it was, like his other educational beliefs, the driving force to how he carried out his professorial duties. He believed that the primary purpose of a university was not to teach the few students who attended, but, through research, to teach everyone. He wrote in 1892: "The function of the University is primarily to teach mankind . . . at all times the greatest men have always held that their primary duty was the discovery of new knowledge, the creation of new ideas for all mankind, and not the instruction of the few who found it convenient to reside in their immediate neighbourhood. . . . Are the Universities to devote the energies of the most advanced intellects of the age to the instruction of the whole nation, or to the instruction of the few whose parents can afford them an—in some places fancy—education that can in the nature of things be only attainable by the rich?"

As can be seen from the quotations we have given from FitzGerald's writing, his interest in education went well beyond the narrow confines of his own department. It was not merely a theoretical interest, for, true to his own beliefs, he took a very practical role in education. He was an examiner in physics at the University of London beginning in 1888, and he served as a Commissioner of National Education in Ireland in 1898, being concerned with reforming primary education in Ireland. As part of this task he travelled to the United States on a fact finding tour in the autumn of 1898.

As one might have expected, his aim was to bring far more practical topics into the syllabus of primary schools. At the time of his death he was involved in the reform of intermediate education in Ireland, and he also served on the Board which was considering technical education.

In 1883 FitzGerald married Harriette Mary Jellett. She was the daughter of the Rev J H Jellett, the Provost of Trinity College and an outstanding scientist, who had been awarded the Royal Medal of the Royal Society of London. It was through his personal friendship with Jellett, and also their joint scientific studies, that FitzGerald got to know Harriette. Although the couple had been married just under 18 years at the time of FitzGerald's death, they had eight children during this time: three sons and five daughters. FitzGerald was elected a Fellow of the Royal Society of London in 1883 and, like his father-in-law, he was to receive its Royal Medal. This was in 1899, when the prestigious award was made to FitzGerald for his contributions to theoretical physics, especially to optics and electrodynamics.

We should now examine the research for which FitzGerald received these honours. Beginning in 1876, before he obtained his Fellowship, FitzGerald began to publish the results of his research. His first work, *On the Equations of Equilibrium of an Elastic Surface*, solved some particular cases of a problem studied by Lagrange. His second paper in the same year was on magnetism, and he then, still in 1876, published *On the Rotation of the Plane of Polarization of Light by Reflection from the Pole of a Magnet* in the *Proceeding of the Royal Society.* He had already begun to contribute to Maxwell's theory, and, as well as theoretical contributions, he was conducting experiments in electromagnetic theory. His first major theoretical contribution was *On the Electromagnetic Theory of the Reflection and Refraction of Light*, which he sent to the Royal Society in October 1878. Maxwell, in reviewing the paper, noted that FitzGerald was developing his ideas in much the same general direction as Lorentz.

At a meeting of the British Association in Southport in 1883, FitzGerald gave a lecture discussing electromagnetic theory. He suggested a method of producing electromagnetic disturbances of comparatively short wavelengths "by utilising the alternating currents produced when an accumulator is discharged through a small resistance. It would be possible to produce waves of as little as 10 metres wavelength or less." So FitzGerald, using his own studies of electrodynamics, suggested in 1883 that an oscillating electric current would produce electromagnetic waves. However, as he later wrote: "I did not see any feasible way of detecting the induced resonance."

In 1888 FitzGerald addressed the Mathematical and Physical Section of the British Association in Bath as its President. He was able to report to British Association that Heinrich Hertz had, earlier that year, verified this experimentally. Hertz had verified that the vibration, reflection and refraction of electromagnetic waves were the same as those of light. In this brilliant lecture, given to a general audience, FitzGerald described how Hertz: 'has observed the interference of electromagnetic waves quite analogous to those of light'.

After his appointment to the Erasmus Smith Chair, FitzGerald continued to produce many innovative ideas, but no major theories. For example, despite his ideas on electromagnetic waves, he had not followed through the research, and the final experimental verification had been achieved by Hertz. The reason for this is perhaps best understood with a quotation from a letter which FitzGerald sent to Heaviside on 4 February 1889: "I admire from a distance those who contain themselves till they worked to the bottom of their results, but as I am not in the very least sensitive to having made mistakes, I rush out with all sorts of crude notions in hope that they may set others thinking and lead to some advance."

Although FitzGerald is modestly talking down his contributions in this quotation, the comment he made about himself is essentially correct. Oliver Lodge gives a similar, but fairer, analysis of FitzGerald's work: "The leisure of long patient analysis was not his, nor did his genius altogether lie in this direction: he was at his best when, under the stimulus of discussion, his mind teemed with brilliant suggestions, some of which he at once proceeded to test by rough quantitative calculation, for which he was an adept in discerning the necessary data. The power of grasping instantly all the bearings of a difficult problem was his to an extraordinary degree.''

Again Heaviside wrote: ''He had, undoubtedly, the quickest and most original brain of anybody. That was a great distinction; but it was, I think, a misfortune as regards his scientific fame. He saw too many openings. His brain was too fertile and inventive. I think it would have been better for him if he had been a little stupid—I mean not so quick and versatile, but more plodding. He would have been better appreciated, save by a few.''

Finally we should examine the contribution for which FitzGerald is universally known today. There had been many attempts to detect the motion of the earth relative to the ether, a medium in space postulated to carry light waves. A A Michelson and E W Morley conducted an accurate experiment to compare the speed of light in the direction of the earth's motion, and the speed of light at right angles to the earth's motion. Despite the difference in relative motion with respect to the ether, the velocity of light was found to be the same. In 1889, two years after the Michelson–Morley experiment, FitzGerald suggested that the shrinking of a body due to motion at speeds close to that of light would account for the result of that experiment. Lodge writes that the idea: ''flashed on him in the writer's study at Liverpool as he was discussing the meaning of the Michelson–Morley experiment''.

In 1895, Lorentz independently gave a much more detailed description of the same idea. It was typical of these two great men that both were more than ready to acknowledge the contribution of the other, but there is little doubt that each had the idea independently of the other. The FitzGerald–Lorentz contraction now plays an important role in relativity.

Sadly FitzGerald died at the age of only 49 years. Maxwell, whose work had proved so fundamental for FitzGerald, had died at the age of 48 while Hertz died at the age of 36. In fact in 1896 FitzGerald had reviewed the publication of Hertz's *Miscellaneous Papers* for *Nature* after Hertz's death. Four years later, in September 1900, FitzGerald began to complain of indigestion and began to have to be careful what he ate. A few weeks later he complained that he was finding it difficult to concentrate on a problem. His health rapidly deteriorated and despite his having an operation the end came quickly.

William Ramsay, on hearing of FitzGerald's death, wrote: ''To me, as to many others, FitzGerald was the truest of true friends; always interested, always sympathetic, always encouraging, whether the matter discussed was a personal one, or one connected with science or with education. And yet I doubt if it were these qualities alone which made his presence so attractive and so inspiring. I think it was the feeling that one was able to converse on equal terms with a man who was so much above the level of one's self, not merely in intellectual qualities of mind, but in every respect.... He had no trace of intellectual pride; he never put himself forward, and had no desire for fame; he was content to do his duty. And he took this to be the task of helping others to do theirs."

FitzGerald was described by Lord Kelvin as: "living in an atmosphere of the highest scientific and intellectual quality, but always a comrade with every fellow-worker of however humble quality.... My scientific sympathy and alliance with him have greatly ripened during the last six or seven years over the undulatory theory of light and the æther theory of electricity and magnetism."

Further Reading

Article by Alfred M Bork in *Dictionary of Scientific Biography* (C C Gillispie, ed) (Scribner, New York, 1970–1980).

The Scientific Writings of the Late George Francis FitzGerald, J Larmor (ed) (Hodges, Dublin, 1902).

George Francis FitzGerald, O J Lodge, Obituaries in: *Nature* (7 March 1901); *Obituary Notices of the Royal Society of London* (1901); *Physical Review* (May 1901).

John Joly

1857–1933

Denis Weaire and Stephen Coonan

In 1885 the magazine *Inventions and Inventors* published a chart of portraits of notable inventors. Included among them was John Joly, the young Assistant to the Professor of Civil Engineering at Trinity College Dublin. He had embarked on the design of a remarkable sequence of physical instruments. From then until the turn of the century he published prolifically on new instrumentation and registered at least 40 patents. The products of his industry and imagination included a telegraphic barometer, a photometer, the meldometer, the aphorometer, a hydrostatic balance, a sea sounder, the steam calorimeter, a solar sextant, an ampere meter, a method of colour photography, sound recording by photography, a rain gauge, an electric furnace, and many more. Even in that energetic age of Kelvin, this was an impressive display of talent and ambition. He must have seemed destined to find his fortune in the commercial world. In fact his most promising business venture, based on a practical system of colour photography, ran into legal contests and never brought a large return—or so he claimed. Instead of continuing along that Edisonian path, he veered off into geology, and became one of the founders of modern geophysics.

Some of young Joly's evident self-confidence may have derived from his remarkable family background. The simple facts of his birth in 1857 in King's County as the third son of a minor clergyman, the Rev John Plunkett Joly, conceal a continental pedigree that included leading servants of the crown in France: his mother had the title of Comtesse de Lusi. She brought him to Dublin when his father died, and there he was educated, first by a tutor and then at the Rathmines School. While he was not academically distinguished at school (nor indeed in his undergraduate study of civil engineering at Trinity, although he did obtain the equivalent of a First), his chemical experiments in his bedroom earned him the early nickname of 'Professor' and were a sign of his independent curiosity about nature's workings. At the University he was to find many like-minded people in the Dublin University Experimental Science Association, which met regularly to discuss exciting developments in science and demonstrate the latest inventions. This

John Joly, from young inventor to grand old man (courtesy of Trinity College, Dublin).

provided the ideal sounding-board for his ideas, most of which were tried out before the Association at an early stage. Its leading member was the redoubtable George Francis FitzGerald*; Joly can surely be counted among the many who drew inspiration from that generous source. In 1891 he moved from Engineering, where he had served as Assistant since his graduation, to the position of Assistant to FitzGerald in the Department of Natural and Experimental Philosophy. He must have

The Physical Laboratory, whose realization owes much to John Joly (courtesy of Trinity College, Dublin).

been attracted by its stimulating atmosphere as much as by the consequent pay rise.

FitzGerald was a passionate advocate of both pure and applied science, and an outspoken critic of the College. He died young, and many of his high ambitions remained unfulfilled at the time of his death. In particular the Physical Laboratory, which would provide decent facilities for physics, was no more than a tentative plan. It was to be John Joly who brought it to fruition, a task that called for considerable diplomatic skills. The building, which has been recently named after FitzGerald, was funded by Lord Iveagh. As in today's world, the money was not quite enough to meet the original aspiration. Financial cheese-paring included the reduction of toilet facilities, after checking the established norm with Cambridge University.

In due course Joly also broke (by both advocacy and example) the academic cartel of classics and mathematics, that FitzGerald had railed against repeatedly. These two subjects dominated the College, to such an extent that the Fellowship examination was entirely based on them, to the exclusion of many such as Joly. After conservative forces had relented and waived the examination in the case of Professors, Joly was finally admitted to this privileged inner circle.

So he can be properly regarded as the great man's worthy successor, even though he migrated to another subject. It may be that only his lack of a Fellowship at the time of FitzGerald's death in 1901 prevented his succession to the Erasmus Smith Chair of Natural and Experimental Philosophy.

It was during his period with FitzGerald that he set out to develop a single image procedure for colour photography. In conception his method was not startlingly original, but he displayed great tenacity and technical skill in bringing it to full perfection and practical realization. Much the same can be said of most of his many inventions.

The notion already existed that three monochrome photographs of a subject taken through three colour filters could be used to create a colour image by being projected simultaneously through similar filters. This is a demanding and awkward procedure—how could it be collapsed into something more convenient? Joly's idea was to incorporate all three colours as lines in a single filter. To do so he needed to design an instrument to rule the lines, in a manner probably suggested by the ruling engines used to make diffraction gratings at that time. Much testing of dyes was required to adjust the colours used for this filter, since it was necessary to compensate for the deficiencies of photographic emulsions at that time. In projecting the image a different filter was used, in which the dyes were adjusted to suit the visual response of the eye. This 'Joly Process' must have required long hours of patient work to perfect. Sadly, in most of the surviving slides the dyes (other than red) have altered considerably, so that much of the original effect is lost.

Once the method was ready for commercial exploitation, Joly obtained patents in England and Germany, and applied for patent rights in the United States. A claim that the granting of those rights would interfere with an existing US patent took Joly to America on the first of two occasions. At his first appearance he easily defended his claim. A second challenge was made through a procedure known as 'swearing back' (claiming prior invention, even if not patented), which was permitted under US law. Ultimately the decision went against him, and he had to reveal many details of his process during his defence. Interested competitors were not slow to take advantage.

Joly's colour slides, which still come to light from dusty drawers in Trinity, were mostly of mundane still-life subjects. One of these was a stuffed parrot; having been found in a cleaner's cupboard in Trinity in recent times, the bird itself is a cherished icon of the man and his method.

His monochrome photography enjoyed more freedom: Trevor West reproduced a splendid Joly photograph of athletes clearing a hurdle in *Bold Collegians*, his history of sport at TCD. In 1896 this expertise was applied to the creation of the first X-ray photographs taken in Ireland. They were displayed to his colleagues on 11 February, only a couple of months after Röntgen's discovery. This is typical of the rapid response of Dublin University Experimental Science Association members to hot news from abroad. As elsewhere, the publication of X-ray photographs in the Dublin newspapers caused a great stir.

We can see in his closeted childhood experiments the dedicated laboratory scientist, but some of his interests took him into the wider world and gave notice of his later career. He became an avid collector of minerals and toured the mountains of Switzerland in the manner of Tyndall*. He was also a keen sailor, cruising the coasts of Ireland and Scotland. Many of his inventions reflect those outdoor pursuits, and the friendships that they fostered. The meldometer, for example, is an instrument intended for the identification of small mineral specimens from their melting points. They were also a source for many of his more discursive and speculative theories, ranging from the abundance of life to the canals of Mars. Whenever he moved from narrow technical matters to such grand themes he could command an elegant prose style.

He presented himself as a rather unconventional candidate for the Chair of Geology and Mineralogy when it fell vacant, and was awarded this position in 1897 as the successor to W J Sollas. In his application, having summarized his interests in geology and mineralogy, he declared: "In the union of physics with the branches of science enumerated, I believe a fruitful advance is assured."

An undergraduate magazine warmly welcomed the new professor: "No more popular appointment has been made in College for a long time".

Reviewing the history of geology at Trinity, Gordon Herries Davies declared that Joly 'was arguably Trinity's greatest scientist of the last hundred years'. He classed him as a modern 'equipment-building laboratory geologist', in contrast to the great contemporary 'cross-country-striding' geologists who performed prodigious feats in the field.

Not long after his appointment to the chair of geology, Joly made an estimate of the age of the earth, basing it upon the salt content of the oceans. He arrived at a figure of 100 million years, which he and Rutherford later attempted to reassess in terms of the effects of radioactivity of rocks.

In physics, this was the period in which radioactivity emerged and, in the hands of Rutherford, pointed the way to the new discipline of nuclear physics. Joly was a pioneer of the exploration of the consequences of radioactivity for geology, in the generation of heat and in the traces that it leaves in minerals. One of these, known as the pleochroic halo, is left by the decay of emitted alpha particles. Joly recognized this mechanism and identified the radii of pleochroic haloes with the particular elements responsible for them. In the case of uranium, he saw the signatures of two isotopes.

As in the earlier story of atomic spectroscopy, not all of these data could immediately be accounted for, inviting speculation that some hitherto undiscovered element had been revealed. Joly decided to call it *hibernium*. Alas for his homeland, it turned out to be samarium.

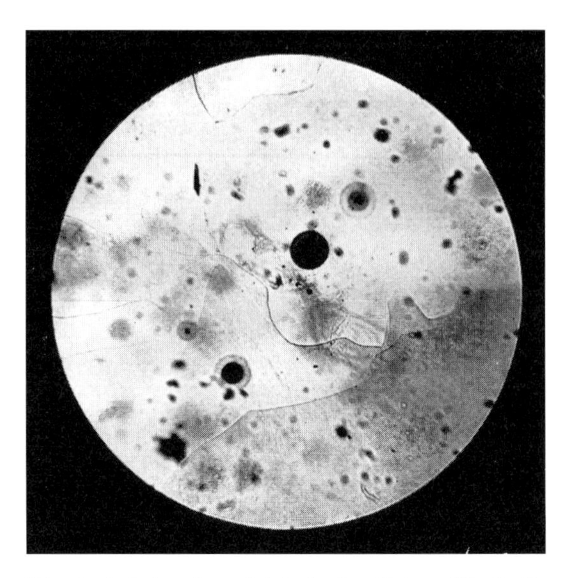

Pleochroic haloes from Joly's *Surface History of the Earth.*

If Ireland was not to have its own chemical element, at least this work drew the admiration of Rutherford, and Ernest Walton* always said that this was crucial in gaining his own admission to the Cavendish Laboratory, upon Joly's recommendation.

Today pleochroic haloes are back in the news. Some of them are invoked by creationists as positive proof of the validity of their own particular view of the age and origin of the earth.

It was at Joly's suggestion that the Radium Institute of the Royal Dublin Society was set up in 1914, and its radiation source used in early applications to cancer treatment. This beneficial application of radium had occurred to him as early as 1903. Characteristically, he invented a form of hollow needle suitable for implanting radioactive isotopes in malignant growths, as the basis for the 'Dublin Method'.

As a confirmed bachelor, Joly was seldom distracted from his deep commitment to his university home, which showed itself in many ways. He fought Trinity's case in the protracted arguments over the Irish University Question. He raised large sums for science from private sources, and would have secured even more from the British government, had not political events overtaken his plans. He showed great concern for students, advocating better arrangements for their accommodation outside the College, and the construction of a swimming pool. These are still topical items for debate today!

He also sallied forth on government commissions and delegations, particularly as Commissioner of Irish Lights, a perquisite often granted

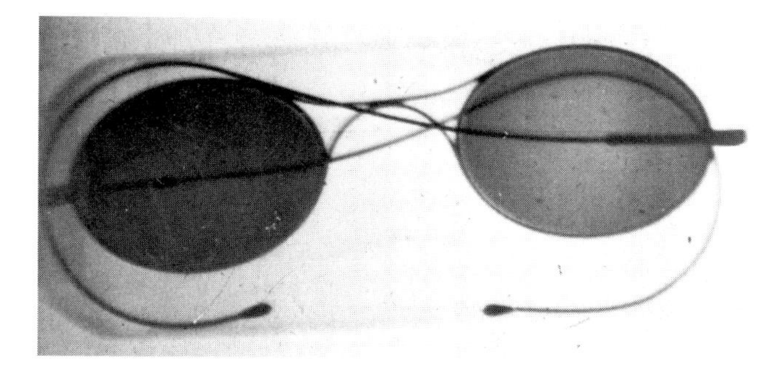

Joly's X-ray picture of his pince-nez spectacles, enclosed in a wooden case (courtesy of Trinity College, Dublin).

to TCD staff. The duties attendant upon this office included congenial sailing tours of lighthouses.

For reasons mentioned above, he 'came in from the cold' to become a Fellow of TCD only in 1918, 26 years after admission to the Royal Society. He led a long and full academic life, producing over 250 papers, four books, and his many inventions. He gave his last lecture six days before his death in 1933. As in so many cases, it must be noted that his later lectures were not so well appreciated by the undergraduates as those given in the first flush of his career.

By then he was a grand old man of science in Ireland, festooned with many honours and distinctions, including Fellowship of the Royal Society, several medals and honorary degrees, and honorary membership of the Russian Academy of Sciences. His contemporaries were amused that this elevated status brought him into contact with Eamon de Valera, now the revered leader of the Irish state. Joly had mounted a guard of volunteers to defend TCD during the Easter rebellion of 1916, making one daring foray into the surrounding chaos in search of intelligence and cigarettes for his men. What did he find to say to his erstwhile foe? Apparently he inquired how de Valera had spent his time in Lincoln Jail, perhaps knowing that the answer lay safely in the study of differential and integral calculus.

As an obituary rightly said of him: "Joly was essentially a physicist, with an interest in everything to which physical principles could be applied". He was also a brilliant maverick (as characterized by Gordon Herries Davies) with more than a touch of the professorial eccentricity that is much prized in certain universities. In his case, the combination of old fashioned spectacles, foreign accent and a motor-bike acquired at the age of 63 created an unforgettable persona. At a personal level, he was a kind and sensitive man. He is justly memorialized in an annual public lecture given in the Geology Department of TCD. Perhaps the name of Joly should also be attached to Trinity's new swimming pool, eagerly awaited at the time of writing.

Acknowledgments
John Nudds did much to preserve the artefacts and memory of Joly during his tenure as curator of the TCD Geology Museum. We are also grateful to its present curator, Patrick Wyse-Jackson, for his advice.

Further Reading

Geology in Trinity College Dublin, Gordon Herries Davies, in: *The Idea of a University*, C H Holland (ed) (Trinity College Dublin Press, 1991).
'The Life and Work of John Joly (1857–1933)', J R Nudds, *Irish Journal of Earth Sciences* **8** (1986) 81.
Recollections and Anticipations, J Joly (Fisher Unwin, London, 1920).
Radioactivity and Geology, J Joly (Constable, London, 1909).
Trinity College Dublin 1592–1952, R B McDowell and D A Webb (Cambridge University Press, 1982).

Joseph Larmor

1857–1942

Raymond Flood

The nineteenth century was an important and notable period in Irish mathematics and science. Ireland produced many fine mathematicians. Some, such as William Rowan Hamilton*, George Francis FitzGerald* and James MacCullagh* spent their working lives there while others such as George Stokes*, William Thomson* (later Lord Kelvin) and Joseph Larmor left the country.

Although Stokes and Kelvin were Irish born, their Irish origins did not seem to influence their science directly, though it did affect their religious and political views. Joseph Larmor was different. As Ivor Grattan-Guinness points out, in his book listed in Further Reading, he maintained an Irish tradition, drawing on Hamilton for mathematical methods, MacCullagh for physics, and working under the influence of his compatriot FitzGerald.

The work for which Larmor is chiefly remembered took place in the decade 1892–1901. This was a transitional period in physics. As A S Eddington argues, at the start of his Obituary Notice of Larmor, the main wave of advance in physics, after half a century of rapid progress, seemed to have run its course, and it seemed that the possibilities for fresh discoveries were limited. Before the end of the decade the discovery of X-rays, electrons and radioactivity had rejuvenated experimental physics. However, at the time when Larmor started his main work there was little to inspire new ideas. Of the researchers who tried to make progress at this difficult time two names stand out—Lorentz and Larmor himself. In Eddington's view: "Their work had much in common, so that it is sometimes difficult to assess their contributions separately. Larmor's reputation has perhaps been overshadowed by that of Lorentz. But on any estimate, Larmor's achievements rank high; and his place in science is secure as one who rekindled the dying embers of the old physics to prepare the advent of the new."

This view of Larmor's importance, not only to science but to the scientific establishment, is echoed by E Cunningham's opening to his obituary of Larmor, also listed in Further Reading: "By the death of Sir Joseph Larmor on 19 May 1942 the scientific world lost one who had

Joseph Larmor (courtesy of AIP Emilio Segrè Visual Archives, E Scott Barr Collection).

made a large contribution to the transition from the classical mechanics to the new physics, and who had served the interests of science and of the larger world as Secretary of the Royal Society and as Member of Parliament for the University of Cambridge.'' Cunningham continues by listing 20 distinctions awarded to Larmor by universities in England, Ireland and Scotland, and by learned societies not only in those countries but also in USA, Bengal, Italy and France.

The man who achieved such national and international recognition was born in Magheragall, County Antrim, on 11 July 1857, eldest son of Hugh, a farmer and then grocer, and his wife Anna. He went to school at the Royal Belfast Academical Institution where he began his distinguished academic career. He is described as a thin and delicate black-haired boy with a precocious ability in both mathematics and classics, eventually obtaining a double first in the scholarship examination at Queen's College, Belfast where he graduated with the highest honours.

From there he went to St John's College, Cambridge, where he lost a year due to illness, but eventually secured first place in the mathematical tripos in 1880. The tripos of that year is especially memorable as the name of Larmor was followed by that of J J Thomson as second wrangler. Thomson, of course, was to become famous as Cavendish Professor at

No. 536.—Vol. XXI. Reg^d at General Post Office as a Newspaper | SATURDAY, MARCH 6, 1880 | WITH EXTRA SUPPLEMENT | Price Sixpence Or by Post Sixpence Halfpenny

TORCHLIGHT PROCESSION OF UNIVERSITY STUDENTS AT BELFAST, IN HONOUR OF MR. LARMOR, THE CAMBRIDGE SENIOR WRANGLER

A torchlight procession at Queen's College, Belfast, honouring Larmor's success as senior wrangler at Cambridge (courtesy of Queen's University, Belfast).

Cambridge, and winner of the Nobel Prize for Physics; with Larmor, he was to lay the foundations of the electromagnetic theory of matter.

Larmor's success led to a fellowship at St John's and his appointment as Professor of Natural Philosophy at Queen's College Galway. He stayed in Galway until 1885, when he returned to St John's to a University lectureship in mathematics. Then in 1903 Sir George Stokes, the Lucasian professor, died. Larmor had built up a high reputation and position in the mathematical community: he had been elected a Fellow of the Royal Society in 1892, was a Secretary to the Society from 1901, and from 1887 had served on the council of the London Mathematical Society. This made him the obvious successor, and he was appointed to hold the famous chair once held by Sir Isaac Newton and subsequently by Stephen Hawking. He retained the Lucasian chair until 1932 when he retired due to ill-health at the age of 75. Larmor had a strong attachment to Ireland, usually spending part of the long vacation there, and it was to Holywood, near Belfast, that he returned on his retirement, dying there in 1942.

Joseph Larmor was awarded many further honours and served in several responsible positions. He was awarded the most prestigious medal of the Royal Society, the Copley Medal, in 1921. In 1914–1915 he was President of the London Mathematical Society. He was knighted in

A sketch of Larmor by M Eug Michel-Maeckler (courtesy of St John's College, Cambridge).

1909, and was Member of Parliament for Cambridge University from 1911 to 1922. Probably his most important book, *Aether and Matter,* was published in 1900, while 1929 saw the publication of his *Mathematical and Physical Papers.*

But what kind of person was he? A modest private man, according to the views of his contemporaries, as publicly expressed. His entry in the *Dictionary of Scientific Biography* by A E Woodruff records that: "Those who knew him report that Larmor was an unassuming, diffident man who did not form close friendships and whose numerous acts of generosity were performed without publicity." He goes on to quote D'Arcy Thompson's telling observation that: "Larmor made few friends, perhaps; but while he lived, and they lived, he lost none."

A S Eddington in his obituary supports this view: "It is difficult to add to this record . . . any intimate details of his ordinary life. He was unassuming and easily approachable; but acquaintances with him never to seemed to grow beyond a particular point. His ready conversation was a screen which seldom betrayed his real thoughts and interests. In some respects he had settled down to be a typical bachelor don. He was

AETHER AND MATTER

A DEVELOPMENT OF THE DYNAMICAL RELATIONS OF THE AETHER TO MATERIAL SYSTEMS

ON THE BASIS OF THE

ATOMIC CONSTITUTION OF MATTER

INCLUDING A DISCUSSION OF THE INFLUENCE OF THE EARTH'S MOTION ON OPTICAL PHENOMENA

BEING AN ADAMS PRIZE ESSAY IN THE UNIVERSITY OF CAMBRIDGE

BY

JOSEPH LARMOR, M.A., F.R.S.

FELLOW OF ST JOHN'S COLLEGE, CAMBRIDGE

CAMBRIDGE
AT THE UNIVERSITY PRESS
1900

[*All Rights reserved*]

Title page of Larmor's book *Aether and Matter*.

often seen walking alone in the Backs and in St John's wilderness; and he evidently loved the charm and quiet of the scene."

But, as is clear from the list of important and responsible positions that he occupied, Larmor appears to have had a strong sense of public duty, both to the mathematical community, and to the Irish question. In his maiden speech in Parliament he took the opportunity to defend the Unionist position in the debate on Irish home rule.

He gathered together his mathematical and physical papers in two volumes which were published by Cambridge University Press in 1929.

They contain 104 articles, 17 appendices, various postscripts and a comprehensive index. His interest in the history of science is explicitly noted in his preface to the first volume where he says, in a typical prolix manner, that "It might have been possible by condensations to reduce the electrical half of this volume by perhaps one quarter of its extent. But, in deference to the advice that was open to him, the writer has abstained from that course. It was urged that out of regard for future historical interests, the order of succession in years should not be blurred. And, moreover, the governing motive for the production of memoirs in formal physical theory, the progressive clarification of the writer's own outlook, may be worth preserving."

Larmor's scientific work centred on electromagnetic theory, optics, mechanics, thermodynamics and geodynamics. His collected works illustrate the development of his interests towards physical optics. But, above all, he had a continuing interest in general dynamics, and in particular the principle of least action. Of the near dozen papers he wrote while in Galway before his return to Cambridge in 1885, the most noteworthy is *Least Action as the Fundamental Formulation in Dynamics and Physics*, which was published in the *Proceedings of the London Mathematical Society*, vol xv in 1884.

In this paper Larmor starts by observing that the relations between problems in different subject areas can frequently best be seen by formulating the problem in terms of obtaining a maximum or minimum as the relationships then are 'easier to grasp and explore'. Hamilton, he points out, had built on the work of Lagrange, to develop a method of reducing the solution of every problem to the consideration of a single function and he is still of the same view on the importance of this approach when, over 35 years later, in an addition to Maxwell's *Matter and Motion*, he writes: "The great *desideratum* for any science is its reduction to the smallest number of dominating principles. This has been effected for dynamical science mainly by Sir William Rowan Hamilton, of Dublin."

Larmor's enthusiasm for the principle of least action, in Eddington's view, was 'intense, almost mystical'. He seemed to require that any theory should be expressed in the form of an action principle. Eddington, for example, writes that he had never been able to persuade Larmor of the truth of general relativity but, in about 1924, he reports Larmor as saying to him 'reproachfully': "I have been reading the continental writers on relativity, and I find it is all least action. I begin to see it now."

The work for which Larmor will be chiefly remembered started as three papers, published between 1894 and 1897 in the *Philosophical Transactions of the Royal Society*, and titled 'A Dynamical theory of the Electric and Luminiferous Medium', subsequently revised and extended in an essay which gained the Adams prize at the University of Cambridge, and then published in 1900 as the book *Aether and Matter: a Development of the Dynamical Relationships of the Aether to Material Systems on the Basis of the Atomic Constitution of Matter, including a Discussion of the Earth's Motion on Optical Phenomena.*

The idea of an all-pervasive ether which penetrated matter and extended throughout the universe was an important concept in nineteenth century science. It was thought that wave phenomena needed such a medium in which to exist and propagate. By 1840 light was generally regarded as a wave phenomenon, which therefore required a mechanical medium subject to dynamical laws. Of the many mechanical models constructed, most postulated the existence of tiny particles with elastic properties whose vibrations carried the wave motion. One of the most successful of these models, consistent with the observed phenomena of light propagation, was developed by the Irishman, James MacCullagh. In 1836 he proposed a medium with a type of elasticity which he called 'rotational', which was easy to describe mathematically, but was quite different from the kind of elasticity shown by any form of matter. Some 40 years later, when it had been shown that light was only one aspect of electromagnetic radiation, G F FitzGerald demonstrated that MacCullagh's model of the ether could also explain the propagation of radiant heat and electromagnetic radiation in general.

In *Aether and Matter*, Larmor developed MacCullagh's ether model by using his preferred technique of least action to explain electric charges. To him the ether was not a material medium and he did not attempt to develop a model working on mechanical lines. Indeed, instead of thinking of the ether as matter he viewed matter as an array of electric particles, 'electrons', which moved about in the ether according to the laws of electromagnetism. However, the electrons were not material particles: "An electric point charge is a nucleus of intrinsic strain in the aether ... if only we are willing to admit that it can move or slip freely through that medium much in the same way that a knot slips along a rope." This radical approach led Sir Horace Lamb, at a meeting of the British Association in 1904, to refer to Larmor's book as being better called *Aether and no Matter*.

In his Preface Larmor explains the role of models of the ether: "The object of a gyrostatic model of the rotational aether is not to represent its actual structure, but to help us realize that the scheme of mathematical relations which defines its activity is a legitimate conception. *Matter may be and likely is a structure in the aether, but certainly aether is not a structure made of matter.*"

An outstanding problem at this time was the effect on electromagnetic radiation of motion through the ether. The ether was thought of as the medium through which light propagated, and was viewed as being like Newton's absolute space—a reference frame against which all other motion could be measured. Light would have a certain speed relative to the ether, and since the earth orbits the sun, it moves through the ether at different speeds and directions at different times of the year. As a result the velocity of light, for example the light coming from a star, as measured on earth, should vary during the year. However, various experiments on different phenomena were unable to determine any difference and Larmor summarizes: "The most varied optical phenomena, whether of ray paths or of refraction, dispersion, interference, diffraction, rotation of plane of polarization, have no relation to the direction of the earth's motion through space, though for many of them the test has been made with great precision."

Larmor went on to discuss the alternative possibility that the earth carries the ether along with it so that the earth is always at rest relative to the ether. However, this is inconsistent with the phenomenon of *stellar aberration*. Aberration, the apparent displacement of a star's position as a result of the earth's motion around the sun, can only be explained if the starlight is propagated through a medium at rest, and the earth is moving through the medium without disturbing it.

To overcome these inconsistent viewpoints (ether carried along with the earth, or the earth moving through the ether) FitzGerald suggested that motion through the ether resulted in a contraction in the dimensions of the moving bodies in just such a way as to conceal the optical effects expected from the motion. This extraordinary suggestion, based neither on theoretical investigations, nor on experimental observations, was surprisingly well received. Larmor's work gave it a new status altogether.

The essential idea was to view the freely mobile intrinsic strains in a rotationally elastic medium (which represent the ultimate particles embedded in the ether) as a system in equilibrium. His approach then was to relate the conditions for equilibrium for different states of motion. The resulting relations or transformations of the space and time coordinates and the components of the electric and magnetic fields of systems moving relative to each other justified FitzGerald's suggestion that the material system must contract in the direction of its motion in order to preserve equilibrium.

These transformations are now named after Lorentz, and the Lorentz transformations are a central component of Einstein's theory of special relativity. Larmor's arguments showed that the transformations were exact to the second order, and it was Lorentz, in 1904, who showed they were exact to all orders. In an Appendix to his Collected Papers Larmor notes the exactness of the transformations, and says it is appropriately named after Lorentz "as having been the initiator of this criterion of invariance... in 1892".

It is uncertain when Larmor himself knew that the transformations were exact. Ebenezer Cunningham, a student of Larmor, wrote: "I did not think he went any further than the second-order invariance of the transformation, nor did I note any suggestion in the book [*Aether and Matter*] that it was possibly an exact transformation.... I left Cambridge for Liverpool in 1904, and there I continued to read and think about the matter and quite on my own I discovered the exactness of the transformation. I wrote to Larmor saying that I had found this to be so, and he replied briefly that he knew this was so, but made no reference to how long he had known it. He certainly had not referred to it in any lecture or publication at that time.... He did not seem at all enthusiastic at the idea that an algebraic transformation happened to be exact. Nor do I recall that he took any particular interest in the speculation that the relativity suggested by the transformation might have far-reaching significance."

This quotation is from the article by Kittel in Further Reading. Kittel discusses why this work of Larmor is so little known today, and concludes that: "Larmor, more strongly than Einstein, was concerned with the interpretation of the ether-drift experiments, and had Larmor full confidence in the consequences of his own transformation, the principle of relativity might have been developed fully some years earlier than transpired."

One result that still carries his name today is *Larmor precession* which is concerned with the motion of charged particles in a magnetic field.

In addition to his own researches Larmor had strong historical interests, editing works of Henry Cavendish, James Thomson, Sir George Stokes and Lord Kelvin and writing valuable biographical notes of various scientists. Noticeable in his longer papers and in appendices to his Collected Works are critical discussions of the work that preceded and led to his own researches.

He was conservative in his scientific views, and it is difficult to say how far he accepted quantum theory and Einstein's theory of gravitation. He is reported as often saying that all true scientific progress ceased about 1900. Eddington writes: "At certain periods at least, his writings on relativity theory were definitely constructive. More usually his references to modern theories give the impression of one who was conscientiously striving to keep an open mind which was not naturally open to the ideas they introduce."

However, he was an active member of the scientific community. As Secretary of the Royal Society from 1901 to 1912, and President of the London Mathematical Society for the two years 1914–1915, and in other positions, he was enthusiastic in promoting new developments and supporting colleagues and students.

Larmor's researches drew on the Scoto-Irish school of physics, which was so important in the mid-nineteenth century, and in particular on work of Hamilton, MacCullagh, Maxwell, Kelvin and FitzGerald. In Woodruff's view: "There is little doubt that he considered himself the last follower of this tradition." Looking back at his work from the beginning of the next century we can see that he formed a vital link between that older nineteenth century tradition and the new physics of Einstein and his followers.

Further Reading

Obituaries by E Cunningham in the *Journal of the London Mathematical Society* **18**(1), January 1943, and by A Eddington in the *Obituary Notices of Fellows of the Royal Society*, No 11, November 1942.

History of the Mathematical Sciences, I Grattan-Guinness (Fontana, London, 1997).

'Larmor and the PreHistory of the Lorentz Transformations', C Kittel, *American Journal of Physics* **42** (1974) 726–729.

Aether and Matter, J Larmor (Cambridge University Press, 1900).

Mathematical and Physical Papers, J Larmor (Cambridge University Press. 1929).

'Electromagnetic Phenomena in a System Moving with Velocity Less than that of Light', H A Lorentz, *Proceedings of the Royal Academy of Amsterdam* **12** (1904) 986.

Matter and Motion, J C Maxwell (reprints: SPCK, London, 1920; Dover, New York, 1952).

Article by A E Woodruff in *Dictionary of Scientific Biography*, C C Gillispie (ed) (Scribner, New York, 1970–1980).

Frederick Thomas Trouton

1863–1922

James G O'Hara

The experimental physicist Frederick Thomas Trouton was born on 24 November 1863 in Dublin, the youngest son of Thomas Trouton. He was educated at Dungannon Royal School, before entering Trinity College, Dublin, where he studied both physics and engineering. He had an outstanding undergraduate career and graduated as first Senior Moderator—which was equivalent to the prestigious Cambridge distinction of becoming a wrangler—in experimental science, receiving the degrees of MA and DSc. He was also awarded the large gold medal in his year, a rare distinction for a science student.

Before graduating he undertook surveying for a railway. He also performed original research in physical chemistry, formulating 'Trouton's law'; this approximate rule expresses a relation between the molecular latent heats and boiling points of various substances, and is equivalent to the statement that the change of entropy per molecule in evaporation at the boiling point is approximately the same for all substances.

Following graduation in 1884 he became Assistant to the Professor of Experimental Physics, George Francis FitzGerald*, with whom he became a cherished friend and colleague. FitzGerald in turn was to be a major influence on Trouton's work in physics. In 1901 Trouton became Lecturer in Experimental Physics at Trinity College Dublin, but the following year he accepted an appointment as Quain Professor of Physics at University College, London. He held this chair until 1914 when disability compelled him to relinquish it; he retained the title of Emeritus Professor thereafter.

Much of Trouton's work was inspired by FitzGerald, for example an investigation to test the accuracy of Ohm's law for electrolytes (1886–1888). FitzGerald was particularly interested in the consequences of James Clerk Maxwell's electromagnetic theory of light, which was published in a series of papers in the 1860s, and then in the celebrated *Treatise on Electricity and Magnetism* in 1873.

When the German physicist Heinrich Hertz began publishing the results of his investigations of electromagnetic-wave radiation—carried out at Karlsruhe between 1887 and 1889—which verified the predictions of Maxwell's theory, FitzGerald and Trouton embarked on a replication,

Frederick Troughton.

extension and interpretation of these experiments. Thus shortly after the publication of Hertz's work in which the finite velocity of propagation of electromagnetic waves in free space and along wires was verified, the Dublin physicists began repeating Hertz's experiments. That was in the autumn of 1888, and in February 1889 Trouton published the results in *Nature* with the title 'Repetition of Hertz's experiments, and Determination of the Direction of the Vibration of Light'.

FitzGerald and Trouton had replicated Hertz's apparatus, which consisted of the world's first radio-frequency signal source—a 6 m dipole vibrator or transmitter—and a resonator or receiver, which consisted of a rectangular or circular wire loop with a spark gap. FitzGerald and Trouton demonstrated that the electromagnetic waves can be analysed into magnetic and electric components and that they are polarized in the same sense as light.

This was followed up by an investigation of the interference of direct electromagnetic radiation with that reflected from a zinc sheet, which was a variation of Hertz's original experiment in which reflection had been obtained from the walls of the lecture theatre at Karlsruhe. Using a Hertzian linear vibrator and a circular resonator, waves of 10 m wavelength were produced, and sparks were observed over a distance of 6 or 7 m. With the resonator in a position where sparking occurred, a large sheet of zinc, 3 m square, was introduced, and placed immediately behind the resonator; the sparking was observed to increase. When the sheet was removed to a distance of 2.5 m behind the resonator, sparking ceased, whereas at a distance of 5 m the sparking was observed again with an intensity slightly greater than in the absence of the sheet. On placing the zinc sheet between vibrator and resonator the sparking ceased completely.

These observations were explained by the interference and reinforcement of a direct 'action' of the vibrator, propagated with finite velocity, and one reflected from the metal sheet. Ordinary masonry walls were found to be transparent to radiation of this wavelength, and in November 1888, at a meeting of the Dublin University Experimental Society, FitzGerald and Trouton demonstrated sparking of the resonator in the College Park while the vibrator was in the laboratory.

In December 1888 they investigated reflection from the surface of a non-conductor. At first no reflection was obtained using large glass plates but reflection from a wall was observed. FitzGerald and Trouton decided now to try Hertz's cylindrical parabolic mirrors which consisted of sheets of zinc nailed on to wooden frames. The vibrator, which produced radiation of about 66 cm wavelength, was placed in the focal line, 12.5 cm from the vertex, and the linear resonator was similarly placed in the focal line of a second mirror, with sparking being observed at a spark gap at the back of the frame. This apparatus made it possible to measure exact angles of incidence. They could observe no effect with glass plates using such mirrors. This they attributed to the thinness of the glass—the reflection from the front surface was thought to interfere with that from the back, the latter having lost half a wavelength in reflection at a surface between a dense and a rare medium.

Following a suggestion made by the Dublin physicist John Joly*, an analogy was drawn with the black spot in Newton's rings, or the darkness seen in thin soap bubbles. With a wall 3 feet thick, reflection was obtained when the vibrator was perpendicular to the plane of reflection but not when turned through 90° so as to be in it. From this observation Trouton drew the principal conclusion of his paper, namely that the magnetic disturbance was in the plane of polarization, while the electric disturbance was at right angles to it.

In addition to Trouton's publication of their results in *Nature*, FitzGerald informed Hertz in a series of personal communications in early 1889. Thus, writing from the School of Engineering at Trinity College Dublin on 23 January 1889, FitzGerald informed Hertz that: "My assistant Mr. Trouton and I have most successfully repeated the experiments on radiation in free space i.e. without a wire. Where our success was imperfect was when we tried to work with shorter wave lengths than about 10 metres and before we had got your paper describing how you had done it. Within the last few days we have been most successful as we could fairly have expected in repeating your recently described experiments. We were successful in repeating your experiment of the interference of a direct and reflected wave that we ventured and succeeded in showing it at a public meeting of the Royal Dublin Society in January. Of course we had to get some of the scientific men present to come down and state to the meeting when they saw the sparks and when they did not, because the sparks were too small for a whole audience to see.

"I expect it is nearly the first time your experiments have been shown in public and certainly the first time in the United Kingdom. This afternoon Mr. Trouton and I have been working at parabolic mirrors and on imitations of your recent apparatus and with no great success as could have been expected on a first trial. We have obtained sparks in the secondary at about 2 meters from the primary and I expect we shall repeat your 20 meters yet. Our parabolic mirrors are not nearly so long as yours and I cannot see why they should be nearly 2 meters long when your receiver is only one meter: I should have thought a little longer than your receiver must be sufficient."

Writing again two days later, on 25 January 1889, FitzGerald informs his German colleague that: "Since last writing Mr. Trouton has succeeded in observing easily at about 4 cm distances and got such strong sparks he was sure he could see them any distance [sic] in the room but by the time he had everything arranged for a more distant observation the knobs had got dirty or something interfered and his more distant observation failed and we have been too busy with examinations since to make any more experiments."

FitzGerald continued his account in a further letter written in London on 8 February, 1889: "The mirrors have proved large enough to observe the reflection from a stone wall 3 feet thick which we long ago observed to be quite transparent and in fact showed to be such at a public meeting of the Exp. Science Association of Trinity College last November. Mr. Trouton set up the experiment and has verified that the radiations are polarised by reflection and that the magnetic disturbance is in the plane of polarisation, as it ought to be on Maxwell's theory. We tried several times for reflection off large sheets of glass with no result and at last Mr. Joly... remarked that we are trying to see the black spot in Newton's rings. We have observed what I hope will be verified as proving some more of Newton's rings in the case of our 3 foot wall. We are getting a pitch tank constructed to make experiments with.

"Have you made any determinations of refractive index by observing the shortening of the wave length when a wall is interposed between a metallic reflector and the receiving circuit as in your original experiments?

The experiment is so easy and obvious; we have not had time to try it yet as we were trying to fix the plane of polarisation which was so much more important. Mr. Trouton was talking of trying it today or tomorrow but as I have had to come up to London I cannot be there to look after it myself".

FitzGerald and Trouton continued their investigations and the experiments undertaken were described by Trouton in a paper in *Nature* on 22 August 1889 with the title 'Experiments on Electro-Magnetic Radiation, including some on the Phase of Secondary Waves'. Continuing the Newton's rings idea, that is to say, the non-reflection of electromagnetic waves from sheets of glass, the following strategy was followed. If the reflecting plate were to be made of such a thickness that the wave reflected from the back had to travel half a wavelength more than that from the front, then the two reflected waves ought to differ by a whole period, half arising from the path difference and half from the change of phase on reflection at the interface between the dense and rare medium. On the other hand, if the path length were to be doubled then the reflected waves ought to differ by half a period and interference would be observed.

For this experiment a large quantity of solid paraffin slabs was used to build up a vertical wall of any desired thickness. Mirrors for transmitting and receiving the radiation were carefully suspended above the wall. Taking 1.75 for the index of refraction of the paraffin, best results were obtained with 68 cm waves and an angle of incidence of 25°; for a wall thickness of 10 cm continuous sparking was observed, whereas for 20 cm only occasional sparking was seen.

FitzGerald and Trouton likewise investigated the phenomena of metallic reflection and multiple resonance, by which the vibrator or primary conductor was found to transmit not a single period of vibration but rather all the values within a range or band.

What then was the value of FitzGerald's and Trouton's investigation of electromagnetic-wave radiation and in particular reflection from substances such as metals, glass and paraffin in 1888 and 1889? In verifying and extending Hertz's investigations of electromagnetic-wave radiation they succeeded in deciding the relation between the direction of vibration in the wave front and the plane of polarization. Trouton also demonstrated a quarter-wave advance when electromagnetic waves are reflected from a small reflector; this was the analogue of a quarter-wave advance introduced earlier by George Gabriel Stokes in interpreting Fresnel's optical investigations. Likewise the experiment carried out in 1834 by Humphrey Lloyd* at Trinity College Dublin was the direct analogue of the Hertzian experiment of the interference of direct and reflected radiation. (See the article on Humphrey Lloyd.) In demonstrating—under the guidance of FitzGerald—such analogies between electromagnetic and optical phenomena, Trouton made a major contribution to establishing the common electromagnetic nature of light and Hertzian waves.

A series of investigations on the relative motion of the earth and the hypothetical ether provides another instance of FitzGerald's influence on Trouton. Considering a charged condenser, moving through the ether with plates parallel to the direction of motion, to possess magnetic energy like a current element, FitzGerald suggested this arose from a mechanical drag on the condenser during charging. Trouton arranged an experimental test but with a negative result. FitzGerald, shortly before his death in 1901, suggested that such a null result might be explained in the same way as the Michelson and Morley experiment, that is, by the hypothesis of a contraction in the linear dimensions of the moving system.

Investigating the problem further, Trouton was led to the view that the experiment involved a turning impulse rather than a translational one. In the event of a positive result, he conceived the possibility of a machine operating by means of the torque derived from air-condensers moving through the ether and thus a new source for power generation. At London, Trouton carried out a very complete investigation on this chimerical turning moment in collaboration with his research student H R Noble but again with a negative result.

Later he devised yet another crucial experiment to discover this effect; this consisted of comparing the electrical resistance of a wire when moving in and across the hypothetical ether stream. This work, carried out in collaboration with A O Rankine, likewise led to a negative result. The null results of these ether-drift experiments, though a personal disappointment for Trouton, were of significance for the further progress of physics; they cast serious doubts on the very existence of an ether and helped prepare the ground for the special theory of relativity.

In collaboration with E S Andrews, Trouton investigated the viscosity of quasi-solids and made a test of Stokes' theory following, by means of X-rays striking a fluorescent screen, the rate of fall of a ball-bearing through a block of wax, which was of the order of 0.9 cm per week. Trouton was the originator of this method for the examination of opaque media.

During his Dublin period Trouton had commenced research on the adsorption of water vapour by various substances in order to construct a recording hygrometer based either on change of weight or of electrical resistance. Later in London he followed up this work and undertook, again with the assistance of collaborators, a long series of investigations on adsorption.

In all Trouton published about 40 papers—a number of them jointly with collaborators—in some 12 journals. In 1897 he was elected a Fellow of the Royal Society, and in 1914 was selected as President of the Mathematical and Physical Section for the Australian meeting of the British Association for the Advancement of Science; serious illness, however, prevented him travelling and delivering his presidential address.

In 1887 Trouton married Annie, daughter of George Fowler of Liverpool, with whom he had four sons and three daughters. His two eldest boys—Eric, a student of physics, and Desmond, an engineering student—were killed in the Great War. In Ireland he resided at Killiney, County Dublin, and in London at Hampstead; in retirement he lived at Tilford in Surrey and later at Downe in Kent.

A severe illness suffered in 1912 led to prolonged prostration; an operation in 1914 resulted in paralysis of the lower limbs and total invalidity but he retained his mental faculties unimpaired. He died peacefully if not unexpectedly at his house at Downe on 21 September 1922; his wife and five children survived him. Contemporaries and historians have noted and discussed his Protestant ethic, his integrity, lively imagination, friendliness and helpfulness, but a certain whimsicality has also been recorded as a character trait. Most importantly, in physics he has been attributed with particular insight into the provisional character of theory and recognition of the importance of experimental research.

Further Reading

Dictionary of National Biography and the forthcoming *New Dictionary of National Biography*.

'Frederick Thomas Trouton—1863–1922, A.W. Porter', *Proceedings of the Royal Society of London* **110A** (1926) iv–ix.

'Prof. F.T. Trouton, F.R.S', E N Da C A[ndrade], *Nature* **110** (1922) 490—491.

'Trouton, Fredrick Thomas, 1863–1922', E Scott Barr, *American Journal of Physics* **31** (1963) 85-86.

Hertz and the Maxwellians. A Study and Documentation of the Discovery of Electromagnetic Wave Radiation, J G O'Hara and W Pricha (Peter Peregrinus Ltd in association with the Science Museum, London 1987).

John Sealy Edward Townsend

1868–1957

Mark McCartney

John Sealy Edward Townsend was born on 7 June 1868 in Galway, the second son of Edward Townsend, Professor of Civil Engineering at Queen's College Galway (now the National University of Ireland, Galway). He pursued his studies at Trinity College Dublin and then at the Cavendish Laboratory in Cambridge, before being elected to the Wykeham Chair of Experimental Physics at Oxford in 1900. He was elected a Fellow of the Royal Society in 1903, and knighted in 1941, the same year as his retirement from the Wykeham Chair.

Townsend's research centred on the behaviour of ions and electrons in gases. He gave the first experimental determination of the electronic charge, and made detailed studies of the conduction of electricity through gases. He is chiefly remembered for his discovery, independently of Ramsauer, of the Ramsauer-Townsend effect, whereby noble gases are effectively transparent to low energy electrons. Townsend died in Oxford on 16 February 1957 aged 88.

Townsend entered Trinity College, Dublin, in June 1885 as a 'pensioner' (that is to say, he did not receive a scholarship) reading mathematics, mathematical physics and experimental science. After a year, however, his natural abilities and hard work were rewarded with a Foundation Science Scholarship in mathematics, and in 1890 he obtained a double Senior Moderatorship (effectively a double first), and gained the highest marks in mathematics for his year. He remained at Trinity until 1895, making a total of four attempts at the examinations to obtain a college Fellowship. After his fourth failure he went to the Cavendish Laboratory in Cambridge to work under J J Thomson as a research student.

In October 1895 Cambridge University introduced a new set of regulations allowing graduates of other universities to gain a Cambridge degree after two years research. To quote J J Thomson, the university's new policy had a 'very auspicious start' with the arrival of the first students. The first was one Ernest Rutherford from Canterbury College, New Zealand, and within the hour John Townsend arrived from Dublin, followed by J A McClelland* from Queen's College Galway.

J.S. Townsend

John S E Townsend (courtesy of Clarendon Laboratory Archives, Oxford University).

J J Thomson's research students at the Cavendish Laboratory, Cambridge 1898. Left to right, (*Back row*): S W Richardson, J Henry. (*Middle row*): E B H Wade, G A Shakespear, C T R Wilson, E Rutherford, W Craig-Henderson, J H Vincent, G B Bryan. (*Front row*): J McClelland, C Child, P Langevin, J J Thomson, J Zeleny, R S Willows, H A Wilson, J S E Townsend (courtesy of Cavendish Laboratory, Cambridge).

Townsend and Rutherford quickly struck up a friendship. On 8 December 1895, Rutherford wrote to his fiancée Mary Newton saying: "Townsend, my particular friend at present, is going up to his cousin's at Yorkshire for Xmas.... I don't know whether I have described him to you. Imagine a middle-sized man, very fair hair rather scanty on top, very fair moustache and a true Irish complexion and a merry blue eye—rather good features and a very pleasant appearance altogether. He is a very fine mathematician and is a good deal of assistance to me in that way. I think it probable he and I will research on abstruse subjects next term, for in some of the work I am doing, it is very difficult to do all without assistance." The work Rutherford mentions was his study of radio waves. Townsend (and to a lesser extent McClelland) helped Rutherford test his transmitter and receiver over various terrain; on the common at night, and from the Cavendish to Townsend's lodgings during the day.

March 1896 saw a 'Science Conversazione' to celebrate the opening of the new buildings at the Cavendish. The great and the good of Cambridge came to socialize and see demonstrations of the various experiments in the lab. Among the demonstrations was: 'Number 30: Mr. E. Rutherford (Trinity). Experiments with Hertzian Waves.' Rutherford's able assistant on the night was Townsend, with them each taking turns at explaining and working the apparatus. Among those who came to view the demonstration was George Stokes* and Sir Robert Ball.

Sir Robert Ball had only recently arrived at Cambridge himself to take the Lowndean Chair of Astronomy in 1893. Previous to this he had been Andrews Professor of Astronomy at Trinity College Dublin and Astronomer Royal for Ireland. He had a reputation for making old contacts from Dublin welcome in his Cambridge home, and hence had already had contact with Townsend (and thus Rutherford). Ball was extremely impressed with Rutherford's work and it was through him, thought Rutherford, that 'the fame of Townsend and me has travelled to all the colleges, so that they are all opening their doors in their anxiety to welcome the research student of whom we are the first examples.' When Rutherford used the phrase 'opening their doors' he meant it quite literally as he and Townsend found themselves invited to dine at table in a number of colleges.

It should not be thought from this that Townsend was in some sense Rutherford's 'side-kick', for Townsend had already by this stage distinguished himself sufficiently as a researcher for J J Thomson to make him a demonstrator in the Cavendish by the end of his first term. Townsend's first research project was on magnetization in liquids, but by the end of May 1896 he had started experimenting on the electrical properties of gases, a topic which occupied much of the rest of his life.

His most important piece of work while at Cambridge was the first determination of the electronic charge, *e*. It was well known by the end of the nineteenth century that gases produced by electrolysis carried an electric charge. By bubbling gas liberated in electrolysis through water, Townsend found that the charged gas caused the spontaneous formation of a fine mist of water droplets, the charges acting as nuclei on which water vapour could condense. Further, by then bubbling this mist through sulphuric acid, the gas could be dried without removing the charge.

Armed with these simple facts, and three pieces of apparatus—an electrometer, a laboratory balance and a camera—Townsend performed the following elegant experiment. First a charged mist is created. The rate of fall of this mist is calculated from photographs. Using this rate of fall Stokes' Law can be used to calculate the average mass of an individual droplet of water. The total mass of water in the mist can be calculated by bubbling the mist through a series of tubes containing sulphuric acid, the resultant increase in mass of the tubes giving the mass of the water. Total mass of the water divided by average mass per droplet gives the total number of droplets. Finally the total charge of the mist is found by driving the dried gas into a collector attached to an electrometer. Assuming that at the centre of each drop of mist there is a single charge the total charge divided by the total number of droplets gives the unit of charge.

Townsend found the result to be independent of the type of gas released from the electrolysis, and to have the value 1.7×10^{-19} coulombs (which compares favourably with the modern value of 1.6×10^{-19} coulombs).

Townsend's supervisor, J J Thomson, used Townsend's ideas to calculate the electronic charge in a slightly different way. He used not an ionized gas from an electrolysis experiment, but the ionized gas produced by exposure of a gas to X-rays, and it is Thomson, and not Townsend, who gets credited with the experimental determination of the value of e. Although Thomson's method was undoubtedly more robust, Townsend's contribution, which formed not only the basis for Thomson's work, but also the much later and more accurate work of Millikan, has been largely forgotten in the shifting sands of the history of physics. Of course the matter wasn't helped by the fact that J J (as he was commonly known to students) had a bit of reputation for using other people's ideas without giving due credit. In later life Townsend would reprimand his children by accusing them of 'doing a J J' if they were naughty.

Townsend's good work in the Cavendish brought its rewards. In 1898 he was made Clerk Maxwell Scholar and in 1899 he was elected a Fellow of Trinity College, Cambridge. In the mean time, like any young researcher, he sometimes became disillusioned. He wrote to Rutherford, who was now Professor at McGill University in Canada: "I must confess that eternally working at ions is beginning to be tiresome." And, like any young researcher, he was also looking out for a permanent academic post. For advice and hints he consulted J J, and wrote to his old professor at Trinity College Dublin, George Francis FitzGerald*. Possible jobs, of varying prestige, in Dublin, Montreal, Bangor (in Wales), and Glasgow were weighed up, but the post he finally got in 1900 was the newly created Wykeham Chair of Experimental Physics at Oxford.

Townsend's fellow professor at Oxford was Ralph Bellamy Clifton, who had been appointed to the Chair of Experimental Philosophy in 1865. His only contribution to science at Oxford was to oversee the building of the Clarendon laboratory, a task which he oversaw with meticulous care, even down to the matter of which instruments to buy. His love for his creation was so great that undergraduates were not allowed to disturb its grandeur by using the apparatus lest they break it. It is therefore perhaps not surprising that although Clifton was happy to have Townsend share the teaching load, he was not happy to have him share the Clarendon. Initially Townsend used some rooms in the University Observatory. By 1902 he had what Clifton referred to as spacious rooms, but which was in fact a large tin hut. But by 1910 Townsend had his own two storey stone built Electrical Laboratory, which among other things boasted a lecture theatre which could seat 100 people.

Around 1900 Townsend made the intellectual leap that the ions in a gas could be considered as another gas mixed with the ambient neutral gas molecules. As such they could be described by Maxwell–Boltzmann statistics, and be expected to exhibit phenomena such as diffusion, which were well-known for ordinary gases. To quote Llewellyn-Jones: "The adoption of this view was a step forward of fundamental importance, as it finally led to the detailed mathematical treatment of the collisions of electrons, ions and molecules in weak fields."

Also around the time of his move to Oxford, Townsend began to study ionization by collision. To understand the importance of Townsend's work, we must first realize the level of confusion present at the time. Men like Thomson and Rutherford had observed that the increase in ionization in a gas in an electric field between two parallel plates, depended on factors such as gas pressure, electric field and plate geometry. Attempts to understand the phenomena were based on the supposed existence of so-called 'double layers' of electricity on the electrodes. Applying an electric field was thought to 'peel off' layers and liberate new charges. Further it was thought that charges were kept in molecules by the pressure of collisions with neighbouring molecules. Thus reducing the gas pressure caused an increased likelihood that charges would 'leak out', hence explaining the fact that ionization rates increased as gas pressure was lowered. Combine this with the fact that physicists generally thought that ionization potentials were an order of magnitude higher than they actually were, and you have some idea of the disarray within the ranks.

Townsend's breakthrough was based on his own experiments, and those of Alexander Stoletov, which showed that the ionization potential of molecules was indeed much less than what was commonly supposed. Since in an electric field the charged ions would gain energy, they could collide with neutral molecules and, since the ionization potential was small, knock out an electron fairly easily. This collision produced two new charged particles which could then in turn gain energy from the electric field, and cause new ionizing collisions.

Townsend realized that the relatively large mass of the molecular ion meant that it was the electron which was the most effective ionizer. Further, given that the energy gained by the electron would be proportional to its mean free path between collisions, which is inversely proportional to gas pressure, and also proportional to the electric field intensity, the experimental data could all be simply explained without recourse to 'double layers' or leaking charges.

Townsend went on to find the coefficient of multiplication (later known as Townsend's first ionization coefficient) which gives the number of ion pairs produced when one electron moves through 1 cm in the direction of the electric field. Later he investigated the role of positive ions in ionization, and introduced what is known as Townsend's second ionization coefficient, which gives the number of ion pairs produced when one *ion* moves through 1 cm in the direction of the electric field.

By the time of the outbreak of World War One Townsend had developed a strong school of research. He had been elected a Fellow of the Royal Society in 1903, and his new electrical lab stood in flourishing contrast to the mausoleum which was the Clarendon under its mortician, Clifton.

On the outbreak of war, Townsend initially volunteered to join a wireless unit which was to be sent to Russia to help train Russian cadets. It is alleged that one of Townsend's colleagues from the Electrical Laboratory called at his home shortly before his departure, and, finding the Professor out, left a message asking Townsend to be sure to write a note to the University recommending the caller as his successor to the Wykeham Chair in the event of Townsend not returning. History does not record Townsend's response.

In the end however the Russian mission did not happen, and Townsend ended up working for the Royal Naval Air service at Woolwich on wireless research. He wrote to Rutherford saying, "enjoying myself hugely with wireless. The practical dodges to avoid jamming and all sorts of tricks are most interesting, so much so that I think I will take up seriously some time [sic]."

In 1915 Clifton retired. The university held off making an appointment until after the war and in 1919 they appointed the 33-year-old Fredrick Lindemann. Under Lindemann the Clarendon started to develop a range of vigorously pursued research areas such as low temperature physics, atmospheric physics, spectroscopy and applied thermodynamics. After a short honeymoon, the relationship between Townsend and Lindemann, and between the two laboratories which they headed, became strained and bitter.

There are a number of reasons for this. Firstly both Lindemann and Townsend were now competing for the same pot of money and resources from the university, a pot which, before the war, thanks to Clifton's apathy, had resided safely in the larder of the Electrical Laboratory. Secondly, even though Townsend was a great physicist, he was a great *classical* physicist. When quantum mechanics came along in the 1920s Townsend was in his mid-fifties. He was never reconciled to the quantum theory, nor relativity. Lindemann embraced both, and hence the Professors' views of the direction in which their subject should be heading were radically different. Finally there was the simple fact that Townsend had been in Oxford for 20 years while Lindemann was a young and ambitious newcomer.

In the early 1920s, Townsend discovered, independently of C Ramsauer, that the noble gases are effectively transparent to low energy electrons. While studying the motion of low energy electrons in argon, he discovered that the mean free path of the electrons was more than 50 times their value in hydrogen. It appeared that the electrons were passing straight through the argon atoms as if they were not there. The subsequently named Ramsauer–Townsend effect ironically finds its explanation via means which Townsend never accepted: quantum mechanics, and the wave nature of the electron.

Although after the war Townsend continued publishing on average two research papers every year, age, the rise of modern physics, and the rise of the Clarendon were against him. By the mid 1930s, the Electrical Laboratory was no longer a centre for growth in teaching and research. It had become moribund, and Townsend, like Clifton before him, had become the unwitting mortician of Oxford physics.

He still had a small stream of researchers coming through the lab, some of whom (such as van de Graff) went on to fame. But students found him difficult to work with, partly because of his hostility to the modern physics of quantum mechanics and relativity. He was getting out of touch with the research and writings of others, and by the late 1930s his lecturing style had become erratic. At times his lecturing was inspiring, but at times it was dreary, with notes being written not only on the blackboard, but on the desk in front of him. To make matters worse he was teaching subjects, such as electronics, where his knowledge was out of date.

In 1941, aged 73, faced with an ultimatum from a visitatorial board of the university to either retire or be sacked, he retired on the condition that the board's decision would remain confidential. Thus a year which had started with the pomp and circumstance of his knighthood, ended in ignominy. Townsend, however, appears to have recovered from the matter relatively quickly. In the spring of 1943 he spent a short time teaching at Winchester College, a prestigious English public school in Hampshire, and he published books on radio transmission (1943), electrons in gases (1947) and electromagnetic waves (1951). He even attended an international conference, which was something he rarely did while a Professor, on ionization phenomena, which was held in Oxford in 1953.

He died in Oxford on 16 February 1957 aged 88, and was survived by his wife and two sons.

Further Reading

'The Origins of Ionization and Plasma Physics', F Llewellyn-Jones in: *Physicists Look Back*, John Roche (ed) (Adam Hilger, Bristol, 1990).

Science at Oxford 1914-1939, J Morrell (Clarendon Press, Oxford 1997).

'John Sealy Edward Townsend', A Von Engel, *Biographical Memoirs of Fellows of the Royal Society* **3** (1957) 257–272.

John A McClelland

1870–1920

Thomas C O'Connor

John Alexander McClelland was an eminent member of the succession of distinguished physicists who emerged from the northern part of Ireland in the second half of the nineteenth century. He was born on 1 December 1870, the youngest of eleven children of an Ulster farmer, William McClelland and his wife Margeret, of Dunallis, Macosquin, near Coleraine. Family lore has it that 'John was no good at rounding up sheep as his mind was always on other things' and that he 'always had his nose in a book and wouldn't even turn a beast in the yard'. This intellectual tendency, and perhaps increasing family prosperity, enabled him to follow a more academic career path. On 24 August 1885, he entered the upper school of the Coleraine Academical Institution, where he showed promise of great ability.

Like many another Ulster Presbyterian of the time he chose to pursue his third level education at what was then Queen's College Galway, currently the National University of Ireland, Galway. One may wonder why this was so. It is said that, for those with limited means, Galway was an attractive location; the cost of living there was less than in other cities, and, because of the smaller number of students, the competition for scholarships and prizes was less intense. Another possible reason was that the recently appointed Professor of Natural Philosophy in Galway was also a Coleraine man, Alexander Anderson, from Camus, whose academic achievements were probably well known in the area.

At that time the College in Galway, like the other Queen's Colleges and various schools and institutions, was operating under the system of the Royal University of Ireland (RUI) which set the syllabuses for courses, conducted the examinations and awarded the degrees, but was not engaged in teaching. A Professor, unless he was a fellow of the RUI, had little control over the content of his lecture courses or the examinations.

There was a multiplicity of courses, and some professorships covered more than one present-day subject, Natural Philosophy, for example, including Experimental and Mathematical Physics. Inevitably this resulted in a very heavy lecturing load. Students were expected to

John A McClelland.

cover a broad range of subjects in their first year before specializing in either the Arts or Science division of the faculty for the BA degree.

McClelland earned a distinguished academic record at Galway. In the first Arts examination of the RUI in 1891, he obtained second class honours in Latin, French, Mathematics and Experimental Physics and was awarded a second class exhibition of £15. In the second Arts examination he achieved first class honours in Mathematical Physics and Experimental Physics, a second honour in Mathematics, and an exhibition worth £18. In 1893 he took the BA degree of the Royal University, with honours in Mathematical Science and a second class exhibition worth 20 guineas. As the Senior Scholar in Natural Philosophy he would have assisted Professor Anderson with the 439 lectures and 108 Practical Physics lectures which were given to a total of 58 students in the Department in the session 1892–1893, and experienced the 'taste' of research which was very much developed in some of the higher classes. In 1894 he was awarded an MA degree with first class honours in Mathematical Physics and Experimental Physics, and a special prize valued £60 and a gold medal.

In 1894 the Royal University of Ireland instituted Junior Fellowships worth £200 per year for four years, and the one in Natural Philosophy was won in open competition by McClelland. It is interesting to note that the Vice-Chancellor of the Royal University, the Right Hon C T Redington, in reporting this to the fourteenth annual meeting of the University, which took place in St Patrick's Hall, Dublin Castle, in October 1895, said that: "The Senate, to show their appreciation of the thesis which he [McClelland] wrote for the examination, have ordered it to be printed and circulated, and I think that such an exceptional honour must be a source of great satisfaction to the distinguished Professors of the Galway College, and to its most honoured President."

A paper by A Anderson and J A McClelland, 'On the temperature of maximum density of water and its coefficient of expansion in the neighbourhood of this temperature', was subsequently communicated to the Physical Society in London on 25 January 1895. A report of the meeting appeared in *Nature* on 7 February 1895. It described the experiments with a dilatometer containing mercury to compensate for a change in volume of the glass, and also mentioned the doubts raised by some of those present on the accuracy of the calibration of the thermometer used. The paper was not subsequently published.

McClelland was also awarded an '1851 Scholarship', valued at £150 per annum, which he used to continue his studies in Cambridge University under Professor J J Thomson. He arrived in Cambridge at the beginning of 1896 at a very exciting time for research in physics. The Cavendish Laboratory there was one of the leading centres for physics, and was attracting research scholars from around the world to work on subjects such as the discharge of electricity through gases and cathode rays.

In 1895 the University instituted a new degree of BA by research and two advanced students, Ernest Rutherford from New Zealand and J S Townsend* from Galway, were already at work there. McClelland joined Trinity College Cambridge as an advanced student under these new regulations, and among his fellow students were C T R Wilson, from Scotland, and O W Richardson, from Yorkshire, both of whom were later awarded Nobel Prizes in Physics for the pioneering work that they began around this time, on the topics of cloud chambers and thermionics respectively.

On his train journey up to Cambridge, McClelland is said to have read in the newspapers about the discovery of X-rays in Germany by Professor W Röntgen, a topic which quickly caught the popular imagination. In January 1896 J J Thomson found that X-rays caused a charged conductor on which they were incident to lose its charge. McClelland's first work was to assist Thomson in a research project on 'Leakage of electricity through dielectrics traversed by Röntgen rays'. Some of the fundamental phenomena of the passage of electricity through a gas exposed to X-rays were observed; the correct interpretation of such facts as the existence of a saturation current was not given until a paper by Thomson and Rutherford a few months later. The heterogeneous nature of the rays was proved by experiments on their absorption by tinfoil. McClelland went on to investigate the selective absorption of X-rays, and his results were published in a short note read before the Royal Society in June, 1896.

McClelland's next paper, on cathode and Lenard rays, was communicated to the Royal Society early in 1897. He showed that Lenard rays, which were observed to emerge from the back of thin plates bombarded with cathode rays, carried a negative charge. It is important to remember that at this time there was a lot of confusion about the nature of cathode rays; they were known to carry negative charge but the charge and mass of the cathode ray particle had not yet been determined. It was later in this year that J J Thomson announced the specific charge or e/m of the corpuscle of cathode rays, which is now accepted as the electron. Another investigation carried out by McClelland at the Cavendish Laboratory was on the figures produced on a photographic plate by electric discharges.

McClelland's most important work at Cambridge was undoubtedly his pioneering development of a method of determining the electrical mobility of ions in gases, as he was to use this technique in much of his subsequent research career. The method consists essentially in drawing a stream of gas containing the ions through a metal tube with an axial electrode, and studying the relation between the current collected and the voltage applied between the electrodes. His seminal paper 'On the Conductivity of Hot Gases from Flames' appeared in the *Philosophical Magazine* in July 1898, and another, 'On the Conductivity of Gases from an Arc and from Incandescent Metals' was published in the *Cambridge Philosophical Society Proceedings* in December 1899.

The observations were explained in terms of ions, and the number, mobility and recombination of the ions was studied. The mobilities of ions from these sources were found to be considerably less than those of the ions produced by X-rays, which had already been investigated by Rutherford, but much greater than those which Townsend was finding for the carriers of the charges in the conducting gas of an electrolytic cell. Their mobility was found to diminish rapidly with increasing distance from the source as the gas cooled down and condensation occurred. The negative ion was always found to have the greater mobility. Some further research on the action of incandescent metals in producing electrical conductivity in gases was not published until the end of 1901, after his return to Ireland. He was awarded his BA by research in 1897 and subsequently a Cambridge MA.

In September 1900 McClelland succeeded Thomas Preston (1860–1900) from Armagh as Professor of Experimental Physics in University

College Dublin (UCD), an institution that had developed from the Catholic University of Ireland. Its main premises at 85–86 St Stephen's Green had a small shed at the back which served as an undergraduate laboratory for chemistry, and there were similarly cramped and impoverished quarters for physics.

One might wonder why a man of McClelland's ability and prospects would take up such a post. In evidence to the Royal Commission on University Education in Ireland in 1901 he said that these Professorships "are not of great value but still Professorships are not so easily obtained that they can be treated with disrespect." He was also aware that his predecessor, Preston, who had been elected a Fellow of the Royal Society in 1898, had done important research work on the anomalous Zeeman effect, and had written widely acclaimed textbooks such as *The Theory of Heat*.

In 1901 McClelland was elected as a Fellow of the Royal University of Ireland on a salary of £400 per annum. His duties were to examine students in all the University examinations in his subject, and to teach in an institution approved by the Senate. He also had the privilege of making use of the relatively large and well-equipped laboratories, used for the practical examinations of the RUI in Earlsfort Terrace, outside of examination times, but he was not allowed to bring any students into the premises. The RUI also provided some expensive research equipment such as a Rowland diffraction grating spectrometer, which served the pioneering research in spectroscopy by W N Hartley, and later T E Nevin and the school of spectroscopists in UCD throughout the twentieth century.

To the same Royal Commission, he expressed his strong views with regard to technical and scientific education in Ireland. He advocated that the universities be made thoroughly efficient, as schools of research in pure science for its own sake, and that the technical schools should always be looking out for its possible applications. Students should receive a general education in secondary schools, and specialize earlier in their university education. Two, or at most three, years of lectures and text-books should qualify the honours student to begin two years of research for an MA. For all this to occur, the equipment must be amply sufficient and the staff sufficiently large. He compared Cambridge, where members of staff might give three lectures per week, to colleges in Ireland where they gave three lectures a day. How little the ideas and complaints of professors have changed in a hundred years!

A result of the Commission's deliberations was the Irish Universities Act, 1908, which set up the National University of Ireland (NUI) with three constituent Colleges at Dublin, Cork and Galway, and an independent Queen's University in Belfast. The College in Dublin acquired the premises of the Royal University in Earlsfort Terrace, the Physics and

Chemistry Departments moved into the laboratories on the south or Hatch Street side of the site, and student numbers began to grow. Shortly afterwards, plans were drawn up for a new building on the site, and McClelland was involved in planning a new Department of Physics.

By 1916 the front and north wing had been completed but the world war and the fall in the value of sterling made it impossible to continue the original building plans. The Department of Physics occupied the ground floor and half of the first floor on the north wing until it moved to its current premises in Belfield in 1964.

McClelland was a valued member of the Academic Council and Governing Body of UCD, and its representative on the Senate of the NUI from its inception in 1908, until his death in 1920. On all of these bodies, his views were of considerable influence in moulding the character of the College and of the University.

Despite the heavy demands on his time for teaching, administration and developing the department, he was active in research. His first publication from Dublin, on 'Ionization in Atmospheric Air', in the *Transactions of the Royal Dublin Society* in 1903, was significant, as it marked the beginning of the researches of the 'Dublin School' on atmospheric electricity and aerosols. However, most of his research in his first decade in Dublin was on the ionizing radiations from radioactive material. In 1904 he showed that gamma rays did not carry negative charges as some had asserted, and that the emanation from radium was also uncharged. He also made extensive studies of the 'secondary' radiation produced when matter is submitted to the action of beta rays.

A series of publications in the *Transactions of the Royal Dublin Society* culminated in a paper on 'Secondary Beta Rays' in the *Proceedings of the Royal Society* in 1908. In most of this work, a beam of beta rays was incident at an angle on a thick plate of material, and the nature and distribution of the radiation from the plate was examined. He showed that it consisted of two parts, the deviation or scattering of the primary beta particles, and also electrons ejected by them from atoms. By using 28 different elements he showed that the secondary radiation increased with atomic weight, although it was not directly proportional to it. This was the first time that a connection was established between atomic weights and the intensity of secondary radiation in the case of beta rays.

A complete interpretation of his results was difficult at that time when the model of the atom in vogue was J J Thomson's 'plum pudding' version. However, as the citation for his election as a Fellow of the Royal Society pointed out: "The line of work opened up in these papers has been taken up by many investigators."

These publications were all under his individual authorship, except for one paper in 1906 on 'Secondary Radiation from Compounds'. This was published in the *Scientific Proceedings of the Royal Dublin Society* with

Felix Hackett (1882–1970), from Omagh, who was a Junior Fellow of the RUI and so had access to its laboratories.

With the establishment of the NUI in December 1908, and the acquisition by UCD of the premises of the now defunct RUI in Earlsfort Terrace, McClelland was able to accommodate research by graduate students. Among the first of these were Henry Kennedy, John J Nolan (1888–1952) from Omagh, and later his brother Patrick J Nolan (1894–1984). In 1912 he published two papers with J J Nolan in the *Proceedings of the Royal Irish Academy* (cited subsequently in *La Radium*) on 'The Electric Charge on Rain', and another with H Kennedy on 'The Large Ions in the Atmosphere'.

These investigations set in train a line of research on atmospheric electricity and aerosols that was continued by the Nolan brothers, and had a profound influence on physics research in Ireland in the twentieth century. For a period from 1934 to 1973 the Chairs of Experimental Physics in the NUI Constituent Colleges at Cork, Dublin and Galway, and also the Recognized College at Maynooth, were held by persons who had done their initial postgraduate research in this branch of Physics.

McClelland was also interested in practical aspects of physics with papers such as 'Comparison of Capacities: an Application of Radioactive Substances' and on 'Electrical Conductivity of Powders in Thin Layers' (with J J Dowling). He was always interested in the application of science in industry. His work as a scientific advisor to the UK Government during World War One may have stimulated his paper (with the Jesuit Father H V Gill) on 'The Self-Ignition of Ether–Air Mixtures', which was undertaken for Nobel's Explosives Company.

McClelland's research work was widely recognized and rewarded. In 1906 he was conferred with a DSc Honoris Causa by the Royal University of Ireland. The Chancellor remarked on the occasion that it was only the second time that the University had conferred an honorary degree on one of its own graduates for distinguished original research. In 1909 he was admitted as a Fellow of the Royal Society. He received an Honorary Doctorate in Science from the University of Dublin in 1917, and in 1918 the Royal Dublin Society awarded him the Boyle Medal, its highest honour for scientific achievement.

McClelland did much useful work outside his College and University, as he had many other spheres of activity. He was Secretary of the Royal Irish Academy from 1906, and a member of the Board of Commissioners of National Education in Ireland from 1910, and of the Board of Technical Instruction in the Department of Agriculture and Technical Instruction for Ireland from 1909. He was a member of the Council of the Royal Dublin Society, and a governor of St Andrew's College for boys in Dublin. In spite of the call of all of these activities on his energies and attention, he undertook more onerous duties as a contribution to the national effort during World War One.

In July 1915 he was appointed to the first Advisory Council of the Committee of the Privy Council for Scientific and Industrial Research. Their brief was to "report and make recommendations on proposals (i) for instituting specific researches; (ii) for establishing or developing special institutions or departments of existing institutions for the scientific study of problems affecting particular industries and trades; and (iii) for the establishment and award of Research Studentships and Fellowships." This was a belated response from the UK Government to the need to foster scientific research, especially by industry, to make up for the many deficiencies in their technology exposed by the war. It was also interested in establishing a more permanent framework for the scientific support for industry in the battle for trade when peace was restored.

McClelland's strong views on science as a support for industry were already evident in his submission to the Royal Commission on University Education in Ireland in 1901, quoted above, and his contributions as a member of the Board of Technical Instruction. A flavour of his sentiments may be gathered from some remarks that he made at the presentation of prizes at St Andrew's College at Christmas 1917. Here he advocated "the necessity of study and research in scientific pursuits—duties incumbent on us if we are to maintain our supremacy in the commercial world and establish our hold in the manufacturing one—in fact, equip ourselves for victory in the industrial war which must occur when blood-fought victories have purchased a lasting peace".

The Advisory Council set about its task energetically interviewing representatives of the scientific, engineering and manufacturing societies throughout Great Britain, while McClelland did the same for them in Ireland. It set up Standing Committees for various industries, encouraged the formation of Trade Associations with their own research institutes, and initiated schemes for research fellowships and the support of individual research projects.

All this necessitated frequent journeys between Dublin and London. During the war, apart from the great discomforts of travelling at that time, every crossing of the Irish Sea was a gamble with death. In the year between August 1918 and July 1919 the Advisory Council met in 19 ordinary and three special meetings, which was typical for its schedule. His colleague, A W Conway, Registrar and Professor of Mathematical Physics at UCD, knew his workload well. In an obituary in *Nature* (22 April 1920) he wrote "The constant strain was too much for him, and oftentimes his friends urged him to take a long rest. His sense of duty, however, prevented him from paying attention to his bodily weakness, and when at last the college authorities persuaded him to take a six months' rest, it was too late''. He died in his home on 13 April 1920. The Report of the Advisory Council for the year 1919–1920 states: ''We have to record with sincere regret the loss by death of Professor J A McClelland, FRS, of University College, Dublin. His frequent attendance at our meetings under the difficult and dangerous conditions of war undoubtedly shortened his life. His death has deprived us of a zealous and active representative of this Department in Ireland. Professor Sydney Young, FRS, of Trinity College, Dublin has been appointed in his place.''

John A McClelland was a devoted family man. In 1901 he married Sarah Josephine (Ina) Murdock Esdale, whose father was a builder in the Coleraine area. They had five children: Dorothy (born 1902), Alexander (1905), Alison (1910), Ina Margeret (1912) and John Dermot (1915). The family lived until 1908 in 23 Victoria Road, Rathgar, and then moved to Rostrevor, Orwell Road, Rathgar. He always loved to return to his roots around Coleraine. Family summer holidays were often spent in a house in Esdale Terrace, Portrush, where the Professor could enjoy his favourite recreation on its golf links or cycling around to visit his friends and relations.

His family pursued various careers in engineering and catering. His elder son, Alexander (Lex) took a civil engineering degree from Trinity College Dublin. After a period of involvement with bridges on the Irish Railway system he moved to England, where he worked for and eventually became head of the Safety in Mines Research Board based at the Royal School of Mines in South Kensington, London.

Professor McClelland was a most able, devoted and inspiring teacher. He was very popular with the students, both in the college and at meetings of the athletic clubs, over many of which he presided. He was always an elegant figure, tall, dark and handsome, with a dark moustache and brown eyes. His friends, like Professor C T R Wilson, remembered him as a quiet, strong, kindly, broad-minded man.

The President of University College Dublin, Dr Coffey, in his report for the session 1919–1920, stated: ''For the College and for his colleagues his loss is irreparable. We knew him to be modest in his greatness, unsparing in his services, a true man in every relation, a great and noble soul, whose association as a fellow-worker and friend was an inestimable privilege.'' Contemporary reports of his passing note that: ''A Presbyterian in religion, he was followed to his grave by men of every shade of thought'', and ''His untimely death is deplored by Irishmen, irrespective of creed and class.'' The minutes of the Royal Irish Academy provide perhaps a fitting summary for this short biography: ''The man who accomplished so much died at the comparatively early age of fifty years in the zenith of his powers. It is not too much to say that in our time few have left us who will be more missed and more regretted. Charity towards all men, the outcome of a life spent in the service of truth, pervaded his thoughts and guided his actions. The world is the better of his life, brief though it was."

Further Reading

'Award of the Boyle Medal to Professor John A. McClelland', *Scientific Proceedings of the Royal Dublin Society* **15**(45) (1918) 677–699.
'Academic Publications of J.A. McClelland', *The National University Handbook, 1908–1932*, Rev T Corcoran (ed) (Dublin, 1932), pp 239–240.
'John Alexander McClelland (1870–1920)', Obituary Notices of Fellows Deceased, *Proceedings of the Royal Society, London, Series A*, **106** (1924) v–ix.

Erwin Schrödinger

1887–1961

Andrew Whitaker

Erwin Schrödinger was born in Vienna in 1887, and, by 1927, principally as a result of his celebrated discovery of wave mechanics, he was recognized as one of the most outstanding physicists of his time. In that year he succeeded Max Planck in the prestigious chair of theoretical physics in the University of Berlin and it seemed certain that he would spend the rest of his career there, but the unsettled nature of European politics in the 1930s and 1940s resulted in him instead spending the years 1939 to 1956 in Dublin, where he was Director of the School of Theoretical Physics in the Institute for Advanced Studies from its foundation in 1940.

His spell in Dublin was, in fact, his longest in any one place during his career, and perhaps one of the happiest. Though he did no research of comparable importance to wave mechanics during this period, as well as publishing many highly respectable papers he wrote a number of important short books on a wide range of topics, one of which, *What is Life?* published in 1944, had a major impact on the postwar development of molecular biology. In this essay, the whole of Schrödinger's life is sketched, but special attention is paid to the Dublin years.

Schrödinger's family was in the comfortable upper-middle class of Viennese society, where financial security was often accompanied by an interest in academic learning or artistic endeavour. His father, Rudolf, had inherited from his own father a thriving linoleum business, but his true interest was in botany—he became Vice-President of the Zoological–Botanical Society of Vienna—and landscape art. The interest of Erwin's mother, Georgie, was playing the violin, but her own father, Alexander Bauer, was a chemistry graduate, and a keen experimental researcher until he lost an eye in an explosion. He became Professor of Chemistry in the Technical University, and also an important figure in the flourishing civic, social and cultural life of Vienna, a curator of the Museum of Art and Industry, and a member of the Theatre Commission. He knew many leading actors and directors, and passed his love of the theatre to Erwin, for whom it remained a passionate interest all his life. Alexander's wife was English, and Erwin became well-acquainted with English things visiting his grandmother's family in Leamington Spa.

He was taught first by his father, then by a private tutor. Between the ages of 11 and 19 he studied in the Gymnasium, primarily Latin and Greek, some German language and literature, and a little mathematics and science. He excelled in all subjects, and had time to develop his passion for the theatre. He proceeded to the University of Vienna where, from 1906 to 1910, he took courses in mathematics, and experimental and theoretical physics, with a little meteorology and chemistry. Major influences were Franz Exner and Fritz Hasenöhrl. Exner was an experimentalist, researching on topics including radioactivity and the science of colour, on which Schrödinger would later do theoretical work. Hasenöhrl, appointed Professor of Theoretical Physics in 1907 at the age of 33, inspired Schrödinger with his brilliant lecturing and informal manner.

A considerable number of important physicists had worked in Vienna in the previous 50 years, including Christian Doppler, famous for his work on waves (the Doppler effect), Josef Stefan (Stefan's Law on black body radiation), and Josef Loschmidt, who made the first estimates of the size of the atom. More significant for Schrödinger were the bitter opponents Ludwig Boltzmann and Ernst Mach. Boltzmann, through his work on the kinetic theory of gases, was the champion of atomism, while Mach, a good physicist, but better known as the leading advocate of positivism, was totally opposed to the idea of atoms. When Mach, partially paralysed by a stroke, had to retire in 1901, Boltzmann returned to Vienna to take his Chair, but, troubled by his decline in health, and distressed by attacks on his work by Mach and William Ostwald, he committed suicide in 1906. This was a great shock to Schrödinger, occurring just before he commenced his own studies.

During his early career Schrödinger regarded himself as a follower of the Viennese School of physics, and was influenced by many of those mentioned; it would seem, though, that he had to transcend this influence to make his own major discoveries.

His doctoral dissertation of 1910 was on dielectrics, and, following a year in the army as an officer cadet, he returned as an assistant in experimental physics under Exner. He enjoyed his work supervising the first-year practical class, but his research was largely on the theories of magnetism and dielectrics; he did some experimental work in atmospheric electricity and radioactivity, but became convinced that his future lay in theory. Then, just before the war, he published his first significant paper, solidly in the Boltzmann tradition. It was a mathematical study of the dynamics of a solid considered as an array of atoms interacting by elastic forces.

Schrödinger served in the army for the whole of the First World War. Until 1917 he was attached to a gun battery, times of intense boredom alternating with battles of great savagery. Schrödinger performed with fearlessness and courage. A 1915 citation referred to his work as

Erwin Schrödinger, army volunteer 1911.

first officer at a gun emplacement in the face of recurrent heavy artillery fire; it speaks of calmness and gallantry, his example to the soldiers, and his success in ensuring that the gun emplacement fulfilled its role. At least he survived. Hasenöhrl was not so fortunate; he was killed leading his battalion in an attack in 1915. In spring 1917, Schrödinger was ordered back to Vienna to teach meteorology to anti-aircraft officers and to run an elementary physics laboratory at the University. During the quiet periods of his war service, Schrödinger managed to perform some high-quality research on such topics as Brownian motion, and in the new areas of quantum theory and general relativity. He also read vast amounts of philosophy, including some Eastern philosophies, and thereafter considered himself a philosopher-scientist.

Schrödinger, Privatdozent circa 1914.

The comfortable lifestyle of the Schrödinger family came to an abrupt end after the war. The linoleum business was destroyed by wartime shortages and Rudolf's only income now was from investments. During the blockade which continued through the winter of 1918–1919, there was starvation in Vienna, and often the family ate at a soup kitchen. Inflation was only starting to bite into their savings when Rudolf died in December 1919 of arteriosclerosis. His widow was forced to leave her fine flat, dying herself of breast cancer in 1921, and Erwin was left permanently insecure about financial matters. He had married Annemarie (Anny) Bertel in 1920, and was always concerned about her material fate if she became a widow.

His academic career, though, flourished in these years, his most important work being in colour theory, in which he rapidly became the leading world authority. It was a fully mathematical treatment of such problems as comparing differences in the brightness of different colours, following from the tradition of Mach, since it worked directly from immediate sensory perceptions. In January 1920 he was offered an associate professorship in Vienna, but in order to have sufficient income to marry Anny he preferred to take a position in April of that year as assistant to Max Wien in Jena, a position which was quickly changed to one of Associate Professor. Since this position was temporary, in October he took up a permanent Associate Professorship at the Technical High School in Stuttgart. Then in April 1921 he made yet another move to obtain a full Professorship in Breslau. He was actually already negotiating with the University of Zurich, though, and in October 1921 he became full Professor there, his fourth move in less than two years.

This last position had many immediate attractions—Zurich was indisputably a major university, he was a successor to Einstein, among other leading physicists, and the move was an escape from the economic and political tribulations of the recently defeated nations. Physically Zurich had its attractions, with lake and mountains easily accessible, though in time Schrödinger was to find the rather conventional social fabric stifling.

During the first half of the 1920s Schrödinger's research became centred on the problems of quantum theory, which was the main topic of the day. The work of Max Planck, Albert Einstein, Niels Bohr and many others in the first quarter of the century constituted the so-called old quantum theory, but by the 1920s it was becoming clear that this theory, though successful in many individual problems, was incoherent, rather a muddle of classical ideas and *ad hoc* quantum rules. A new rigorous theory based on consistent axioms was required.

Schrödinger's eventual way forward was signalled by the French physicist Louis de Broglie. By this time it was a well-established, though not well-understood, feature of the quantum theory, that light, which from the early nineteenth century had been thought to be straightforwardly a wave, in some experiments seemed to manifest itself as a particle. As a counter-thrust, de Broglie suggested that objects such as electrons, till now thought to be particles, should, in addition, have a wavelike nature. In 1924, in his PhD thesis, de Broglie produced the expression for what is now called the de Broglie wavelength of the particle: $\lambda = h/p$, where h is Planck's constant, the fundamental constant of quantum theory, and p is the momentum of the particle. The ideas seemed clever but scarcely realistic, and the examiners were undecided whether to award the PhD, but they sought advice from Einstein, who was extremely positive about the idea.

However, the first great advance came from an entirely different quarter. In July 1925, the 24-year-old German physicist, Werner Heisenberg, a protégé of Bohr, developed the so-called matrix mechanics. His method consisted of dropping altogether any attempt at forming a picture or model of the atom, and working purely mathematically, using two-dimensional arrays of numbers (or *matrices*) for each physical quantity. Remarkably the scheme he produced agreed with the old quantum theory in some of its greatest triumphs. It reproduced the energies of the hydrogen atom previously obtained by Bohr, and also these of the simple harmonic oscillator produced by Planck (though, for the latter case, it increased each energy-level by $\frac{1}{2}hf$, where f is the frequency of the oscillator, which did not affect any of Planck's conclusions).

For Bohr, Heisenberg, and their supporters such as Max Born and Wolfgang Pauli, the very lack of any picture became an essential part of the conceptual advance made by matrix mechanics. Schrödinger, however, found the mathematics repugnant and the lack of any physical model totally unsatisfactory. Soon he had an opportunity to improve matters. He was interested in de Broglie's ideas, and in November 1925 he elected to report on them to the Zurich physics colloquium. Rather casually at the end, Pieter Debye mentioned that talk of individual waves was rather childish. Surely, as in other branches of physics, one required a general wave *equation*.

Schrödinger decided to search for one, and was triumphantly successful. Before the end of January 1926, he had written his first great paper, in which he produced a motivation for his wave equation, and showed that it was able to give the energy-levels for the hydrogen atom and the simple harmonic oscillator agreeing with those of Heisenberg. The form of the equation for the one-dimensional case, the famous time-independent Schrödinger equation was

$$-\frac{h^2}{8\pi^2 m}\frac{d^2\Psi}{dx^2} + V\Psi = E\Psi$$

where V is the potential energy, m the mass of the particle, and Ψ the wave-function. For the case where V is zero, we just get the de Broglie solution for a free particle, while, for any other form of potential, we must insert the appropriate form of V into the equation.

During the first half of 1926 Schrödinger published a series of papers surveying in a masterful fashion the whole field of what became known as wave mechanics. He included deep-seated analogies between mechanics and optics based on the work of William Hamilton*, application to the diatomic molecule, the use of perturbation theory, which is an approximate method when the potential is too complicated mathematically for the problem to be solved exactly, and the extension to time-dependent problems. This series of papers must be reckoned as one of the greatest sustained achievements of any branch of science.

What has not been mentioned is Ψ, the subject of the Schrödinger equation, known as the wave-function, and here Schrödinger felt he had achieved a considerable advance on Heisenberg, since he thought that the wave-function gave a real picture of the atomic system. He believed that it showed that reality was fundamentally wavelike, the wave-function representing some sort of real electromagnetic charge density. This became known as the Schrödinger interpretation of quantum theory, completely distinct from the actual mathematics. However, here he was doomed to disappointment. Schrödinger himself was one among a number of physicists who showed that, though matrix mechanics and wave mechanics seemed entirely different, they were actually two different representations of the same underlying mathematical ideas.

Just as Schrödinger had been repelled by the retreat *from* atomic models, so Bohr and Heisenberg were repelled by Schrödinger's attempt to bring them back, in whatever form, and, in long and intense discussion with Bohr, Schrödinger gradually became convinced that his interpretation could not solve many major difficulties. In particular, the wave-function for a system of two particles could not be thought of as two waves in a three-dimensional space, but as a single wave in a six-dimensional space, which suggested that a wave-function must be a mathematical, not a physical, object. From the late 1920s on, the almost universally held view was that of Bohr and Heisenberg—that no physical picture could be obtained for atomic phenomena.

The main opponents of this view were Einstein and Schrödinger. Einstein's attacks centred on the famous EPR paper of 1935. (See the article on John Bell in this volume.) Schrödinger's most famous idea, also postulated in 1935, was the so-called Schrödinger's cat. In this feat of the imagination, a cat is enclosed in a container with a radioactive atom for a time-period T. In this period the probability that the atom will decay is $\frac{1}{2}$; so is the probability it does not decay. If it decays, it is arranged that poison is released and kills the cat. At the end of period T, the atom is in a linear combination of states given by $(1/\sqrt{2})(s+d)$, where s means it has survived, d that it has decayed. The atom is neither in a survived nor in a decayed state, but in a superposition of both. Such is fairly standard in quantum theory, but much more difficult to appreciate is that the cat too is in such a state, represented by $(1/\sqrt{2})(a+d)$ where a represents a cat alive, and d the cat dead. The cat is neither dead nor alive, but in a totally unphysical hybrid state. Only when it is observed, that is to say only when the box is opened, does the cat become wholly dead, or wholly alive. Such is the consequence of the approach to quantum theory of Bohr and Heisenberg, which stipulates that there is nothing that can be said about a system beyond what the wave-function explicitly tells us. Schrödinger did not agree.

Even by the time this paper was written, though, his life had changed dramatically. First, very soon after he produced wave mechanics, he was appointed to the Chair of Theoretical Physics at the University of Berlin on the retirement of Planck. He took up the position in October 1927, and in 1929 was also elected to membership of the

Prussian Academy, both awards putting him at the pinnacle of his profession. In Berlin he gave an immediate impression as an extremely lucid lecturer, informal in his relations to students.

However, with the coming of the Nazis, Schrödinger felt increasingly repelled and depressed. An opportunity to resolve matters came when he met Frederick Lindemann, Professor of Physics at Oxford University, later scientific adviser to Winston Churchill during the Second World War. Lindemann was visiting Germany to try to recruit a few Jewish physicists for Oxford. He had obtained a number of temporary grants from the chemical manufacturers ICI for these people, and he was amazed when Schrödinger, though not Jewish, indicated that he would be interested in going himself.

Matters were fixed, and Schrödinger resigned from Berlin and arrived in Oxford in November 1933; he had been elected a Fellow of Magdalen College, and ICI had put together a substantial financial package for him. Almost simultaneous with the arrival came the news that he was sharing the 1933 Nobel prize for physics with Paul Dirac, who had also made massive contributions to quantum theory. The 1932 prize had been delayed, and was now awarded to Heisenberg. This award further increased Schrödinger's prestige at a very welcome time, but even so, he was unhappy in Oxford. He was asked to do very little lecturing, and generally felt that he was treated like a charity case.

An offer to take up a chair at Edinburgh in Scotland fell through, partly because of delays, partly because Schrödinger felt the attraction of another offer back to Austria to the University of Graz. He took up the latter post in October 1936, but in March 1938 German troops occupied Austria (the Anschluss) and in August he was dismissed. The Schrödingers fled to Rome, where they were contacted by Éamon de Valera, Prime Minister in Dublin. They met him in Geneva, where he was acting as President of the League of Nations.

De Valera, himself a keen mathematician, aimed to set up an Institute for Advanced Studies in Dublin. In the country of Hamilton, he felt that a School of Theoretical Physics would be eminently suitable, as well as comparatively inexpensive, the other foundation School being of Celtic Studies. He was anxious to secure the services of Schrödinger as Director of the School of Theoretical Physics, and this was arranged in Geneva, but while the Institute was being set up, Schrödinger spent a pleasant year at the University of Gent in Belgium.

The family arrived in Dublin in October 1939. It should be mentioned that Schrödinger's ways were not those of the bourgeoisie. He had many amorous liaisons outside marriage. At this stage his family consisted of Erwin and Annie Schrödinger, Hilde March, wife of a colleague of Schrödinger, who was effectively Schrödinger's second wife, and Ruth, the daughter of Schrödinger and Hilde. He was to have

Left to right: Schrödinger, Born, Lonsdale, Irish President and scholar Douglas Hyde (in wheelchair), Ewald, Irish Prime Minister Éamon de Valera in Dublin (courtesy of AIP Emilio Segrè Visual Archives, Ewald Collection).

several other relationships, and father two further children outside marriage during the Dublin years.

In every way he played a full part in Dublin life, particularly in the thriving theatre, but also in many aspects of social and cultural life. Schrödinger called his period in Dublin "a wonderful time"; occasionally they would joke quietly among themselves: "We owe it to our Führer".

During this period, Schrödinger produced much respectable but unremarkable research, but, like Einstein, his main aim during these years was to work towards the great goal of a unified field theory which would unite gravitation and electromagnetism. There were times when Schrödinger thought he was making progress, but eventually both he and Einstein had to admit that the problem was too hard.

At least in retrospect the most significant outcome of Schrödinger's Dublin years was a series of short books based on public lectures. By far the most important and influential of these was the book *What is Life?* based on a course of lectures given in 1943; the lectures themselves were attended by a huge audience including de Valera and many other notables of Dublin society. Schrödinger had been struck by recent work

Schrödinger on the banks of the Liffey.

on the gene, concluding that its stability over many generations is related to it being a molecule. He imagined it a very large protein molecule, though it was at this period that Oswald Avery recognized that it was actually DNA. Mutations occur only rarely, because the process is not continuous but quantum in nature, from one discrete state to another. Schrödinger gave an extensive account of the gene and how it mutates, largely based on the experiments of Hermann Muller and the theories of Max Delbrück.

This book was found immensely stimulating by a number of physicists who were to make enormous contributions to the foundation and rapid progress of the new subject of molecular biology. The idea that quantum theory could be essential in the solution of biological problems was entirely new to most physicists, and in particular Schrödinger raised the possibility of what would be called a *genetic code*, determining the nature of an individual organism. He also suggested that, as well as quantum theory, entirely new physical principles might be required to understand the nature of life. Among physicists influenced to study biological problems were Francis Crick, James Watson and Maurice Wilkins, all subsequently winners of the Nobel prize for their work on the structure of DNA.

Ironically recent studies of *What is Life?*, in particular by Linus Pauling and Max Perutz at the Schrödinger centenary conference in 1987, have suggested that there was little new in Schrödinger's work, which also demonstrated some misunderstandings of both biology and thermodynamics. This does nothing, of course, to diminish its historical importance, or indeed to remove credit from Schrödinger for tackling a new and important area of science in such a stimulating way.

Another set of lectures of great interest is *Mind and Matter*, written by Schrödinger as the 1956 Tarner Lectures, though he was unable to travel to Cambridge to give them, and they were read from his manuscript. Schrödinger described the results of his long-standing concern with the relation between consciousness and the physical universe; his interest in Eastern philosophy had led him to belief in a universal consciousness, to be identified, in fact, with the world picture produced by science, one world crystallizing out of many apparent minds.

Much the same ideas had been the basis for a series of seminars given in the early 1950s in Dublin, in which Schrödinger returned to his prime concern with quantum interpretation, suggesting a central role for mind in quantum theory, in particular in the process of quantum mechanical observation or measurement. These seminars remained unpublished at the time, but in the 1990s the French philosopher-physicist Michel Bitbol published both his edited version of these seminars, and a full study of Schrödinger's philosophy of quantum theory and its historical development. Bitbol suggests that Schrödinger's ideas anticipate in part several of the interpretations of quantum theory that have developed since Schrödinger's time, such as the many worlds and modal interpretations.

By the mid-1950s circumstances were right for Schrödinger to return to Vienna. This would have been possible at the end of the war, but Austria was then occupied by the four powers, and there seemed every likelihood of a Soviet takeover, which Schrödinger could not contemplate. In retrospect he felt his last decade in Ireland had been of great value. But after a 1951 sabbatical in Innsbruck, he felt a great pull to return to Austria, which would also be necessary financially; in Austria he would receive a full pension. By 1955 Austria's neutrality was guaranteed by Moscow, and in 1956 he became Professor in the University of Vienna, retiring in 1958. These next few years were happy ones for him and Anny in their return to the land of the their birth, but their health gradually deteriorated, and Erwin died on 4 January 1961.

Acknowledgment
The author is grateful for a helpful reading of the manuscript by Mrs Ruth Braunizer, daughter of Erwin Schrödinger.

Further Reading

Schrödinger, Life and Thought, Walter Moore (Cambridge University Press, 1989) is the standard extensive biography, much used in the preparation of this article.

What is Life? with *Mind and Matter* and *Autobiographical Sketches,* Erwin Schrödinger (Cambridge University Press, 1992).

Schrödinger, Centenary Celebration of a Polymath, C W Kilminster (ed) (Cambridge University Press, 1987).

Schrödinger's Philosophy of Quantum Mechanics, Michel Bitbol (Kluwer, Dordrecht, 1996); see also *The Interpretation of Quantum Mechanics (Dublin Seminars 1949-55 and other unpublished texts),* E Schrödinger (M Bitbol, ed) (Ox Bow Press, Woodbridge, Connecticut, 1995).

Cornelius Lanczos

1893–1974

Barbara Gellai

Once, in his later years, Cornelius Lanczos was asked what above all he would like to be remembered for. His answer, given in Rodin's article in Further Reading, was clear and definite: "For having introduced a new style of text book". His answer may have been surprising because one might have expected him to have chosen some of his well-known and widely-used methods in applied mathematics, or one of his many papers on the theory of relativity, quantum mechanics or electromagnetism. In this new style of textbook, he put the emphasis on ideas and concepts and their mutual interrelation, rather than on the mere manipulation of formulae.

In addition, his lecture style was captivating for both professionals and laymen. He had the unique ability to present mathematical ideas tailored to the mathematical level of the audience, about which he always inquired previously. Although his English suggested that the language was not his native tongue, he could use this to add very personal touches to his lecture. In 1960 the American Mathematical Society awarded the Chauvenet Prize to him for his paper *Linear Systems in Self-Adjoint Form* (see Further Reading). The prize is awarded annually to the author whose mathematical paper has been judged best from the viewpoint of exposition and lucid presentation.

It was a laborious procedure before he reached the above-mentioned level in the language. In 1932, Lanczos accepted a position at Purdue University, Lafayette, Indiana, in the United States. He soon realized that his career would to a large extent depend on how good a teacher he could become. Overcoming the language problem was one of his greatest challenges. He accepted a position with teaching responsibilities, while his knowledge of English was practically nil.

The first year in his new position was characterized by his heroic effort to master English. One of his colleagues first listened to Lanczos's comments in a private audience, and helped him to prepare his lectures by writing the correct pronunciation below each word. He was teaching six hours per week and each of his lectures required about six hours of preparation.

Cornelius Lanczos was born on 2 February 1893, in Székesfehérvár, Hungary. Lanczos's father was a respected Jewish lawyer and a cultured man. Education was highly valued in the family. Lanczos attended the local Cistercian Catholic gymnasium, a superb senior high school where he graduated in 1910 with an average 'jeles', which is an excellent grade.

In the autumn of 1911, he was admitted to the Faculty of Arts of the University of Budapest where he studied physics, mathematics and philosophy. At that time, Hungary was enjoying one of the most booming periods in its history. The capital, Budapest, was pulsing with mental vigour in its 600 middle-class coffee houses and its elite educational system.

Lanczos attended Lipót Fejér's lectures in differential and integral equations. At the age of 31, Fejér had been appointed Professor of Mathematics at the University of Budapest and was already an internationally known mathematician. Lanczos's teacher in physics was Baron Roland von Eötvös, a former student of Gustav Kirchhoff and Robert Bunsen at the University of Heidelberg. Later he was renowned for his precise measurement of the ratio of gravitational mass and inertial mass.

According to Lanczos, Eötvös's main assistant with classroom experiments was an older individual with no formal education in physics. When Eötvös was asked why he insisted on having an uneducated assistant, he said: "Because he knows no physics whatsoever, he does not dare to change anything with my experiment but does everything the way I said." Eötvös's results later proved to be an experimental justification of the principle of equivalence of Einstein's general theory of relativity.

Because of his natural receptiveness to speculative thinking, Lanczos was impressed by the theory of relativity. However, he did not immediately accept Einstein's theory. In his annual Statutory Lecture on *Albert Einstein – His Life and His Work* in the Dublin Institute for Advanced Studies for the academic year 1955–1956, he said: "In my student days the theory of relativity was in its infancy and [was the] subject of heated controversies. I learned relativity in a half-hearted way and rejected it wholeheartedly. I could not understand how it happened that there was a sphere and yet it was not a sphere. When it was convenient, you had to think of a certain object as a sphere, and then again, when it was convenient otherwise, as an ellipsoid. The whole theory looked to me like a fake, just made up in such a way that one should always get the correct result. Needless to say, the fault was entirely my own and not that of the theory.... My doctoral thesis dealt with a problem of special relativity."

After Lanczos earned his PhD on the topic *The Function Theoretical Relation to the Maxwellian Aether Equation*, he left Hungary for Germany

Lanczos in 1910 after finishing his studies at the Cistercian Gymnasium (courtesy of the late Istvàn Deák Sr.).

in 1922. He wanted to be part of the far-reaching revolution of twentieth-century physics which was taking place at that time in Germany.

Lanczos's first position was at the University of Freiburg from where in 1924 he went on to Frankfurt am Main, where he became the Assistant to Erwin Madelung, Head of the Department of Theoretical Physics. Epoch-making changes were about to begin in quantum mechanics at that time. The papers of the three leading scientists of the field—Werner Heisenberg, Max Born and Pascal Jordan—were published in 1925.

Lanczos recognized that the new quantum mechanics has a deep relationship to the theory of integral equations, which would be in complete harmony with the field-like conception of physics. He transcribed the matrix theoretical methods of Heisenberg–Born–Jordan and submitted his paper on the topic for publication in German on 22 December 1925. One problem, however, still remained unsolved: he needed to find the proper form of a special function in the equations, but Lanczos was not able to figure this out. Two weeks later Schrödinger's* paper on differential equations was published, in which the function got its proper form. It was not the form with cosmic meaning which Lanczos would have liked to find for it, but it was a form which made the theory complete.

Decades later in 1972, in Trieste, Italy, at *The Physicist's Conception of Nature* symposium, B L van der Waerden gave a lecture on Lanczos's paper and discussed its merits in detail. Lanczos appreciated the 'rehabilitation' of his paper, but being a realistic and a humble scientist he gave the credit to Schrödinger with the witty remark to van der Waerden that "...it is the second man who hits the nail on the head and not the first one."

It was not Lanczos's nature to worry too much about a problem after it has been solved properly by someone else, though he himself may have struggled with it too. The whole field of physics—especially the theory of relativity—was waiting for him with its mysteries. In 1928 he accepted the invitation of Albert Einstein to go to Berlin. Einstein desired that Lanczos should apply his mathematical talent to the theory of relativity.

The encounter with the originator of the theory was also not without troubles. Just as Lanczos had difficulty with the theory of relativity, he also had difficulty with Einstein himself. They did not quite agree about the topic Lanczos would be working on. Lanczos had in mind a very specific program, the problem of motion, while Einstein would have liked him to work on a certain development of the theory of relativity. The method Einstein wanted to apply seemed to Lanczos too artificial.

Einstein, who had little interest in ideas which did not fit exactly into his frame of reference, had the impression that Lanczos did not have the necessary 'enthusiasm' for the cause. Perhaps there was some truth to this, since, instead of working on 'distant parallelism' to improve the theory of relativity as Einstein would have liked him to do, Lanczos published a series of four papers on Dirac's equation in which he discussed the fundamental problems of matter fields and the origin of mass.

After one year, their cooperation ended and nothing was ever jointly published. Aside from the purely scientific reasons, there may have been another factor which contributed to the failure of the collaboration. Although Lanczos had deep admiration for Einstein, he needed to develop his own scientific ego. Their cooperation ended, but their correspondence, and occasional personal encounters, lasted until the death of Einstein who, in his later years, highly valued Lanczos's friendship and helped him with personal advice whenever Lanczos's life reached a crossroads.

Around this time, in 1929, Lanczos married a German woman, Maria Rupp. Their marriage soon was disrupted by his wife's contracting tuberculosis, and she was compelled to spend much of her time in Swiss sanatoria.

Because of the adverse political situation in Germany in 1932, Lanczos accepted the position of Professor of Mathematics at Purdue University. Because of her illness, his wife did not get permission to enter into the United States, and Lanczos was compelled to leave her behind with his parents in Székesfehérvár, Hungary. He accepted only a part-time position and every year from 1931 until his wife's death he spent half a year in the United States and travelled back to Hungary to spend the free semester with her, and pursue his research.

In spite of the extraordinary difficulties which came with the several weeks' travel by ship and train every year, the Purdue years were extremely productive ones in Lanczos's life. His talent in applied mathematics (he called it 'workable mathematics') produced two remarkable methods: the τ method and the FFT (fast Fourier transform), a discovery heralded as 'before its time'.

The τ method was an extraordinarily effective approximation method for both empirical and analytical functions. Lanczos preferred to use the terminology 'telescopic method' because just as the telescope excludes every unnecessary detail which prevents focusing on the subject, so his method eliminated everything which prevented convergence to the target function. In the case of certain functions, the method produced an approximation 10^6 times better than did traditional methods.

The other method was what has been called a rediscovery of the FFT. He was approached by one of his academic colleagues to help to solve a problem which needed the Fourier transform. The calculation of the Fourier coefficients was a great problem at that time because of the lack of a feasible algorithm. By using the symmetries of sines and cosines, Lanczos was able to derive an algorithm which later was found to be an equivalent to the core of the Cooley–Tukey FFT algorithm.

In addition to his increasing interest in applied mathematics and his teaching responsibilities, Lanczos was able to publish 11 physics papers between 1931 and 1935. Among the papers was one entitled *Matter Waves and Electricity,* which later, in 1954, attracted Schrödinger's attention. As a result, Schrödinger invited Lanczos to Dublin. Until then, however, he remained in the United States and in 1946 he left Purdue for the Boeing Aircraft Company in Seattle.

The industrial environment made it possible for Lanczos to develop another of his exceptional abilities which made him into what one could call a 'scientific interpreter'. That is, he could translate the problems of engineers into the language of mathematics, and, conversely, reformulate the mathematical results in terms of engineering. At the Boeing Company, Lanczos in his research and consultation encountered problems in the fields of network analysis, airline flutter problems and vibration of antennas. The analyses of these problems required the solution of eigenvalue problems. Lanczos developed an ingenious method for finding eigenvalues with great accuracy by using a method he called 'minimized iterations'. The method made it possible to avoid the accumulation of rounding errors.

He completed his method when, in 1949, he joined the Institute for Numerical Analysis of the National Bureau of Standards in Los Angeles, a

Lanczos at work in the Boeing Aircraft Company, Seattle, 1947 (courtesy of Ida Rhodes).

national centre for mathematical computation. The paper has been published in the *Journal of Research of the National Bureau of Standards* **45**(4) (October 1950) 255–282. The method in this paper is widely-known as the Lanczos method.

Outstanding mathematicians worked on developing the method and adapting it to large matrices where, because of rounding errors, the method lost its accuracy. Recently, *Computing in Science and Engineering* **2** (2000) 1, a joint publication of the IEEE and the American Institute of Physics, designated the Lanczos method, and its later applications, as one of the top ten algorithms of the century.

Around the time he moved to Los Angeles, his first textbook entitled *Variational Principles of Mechanics* was published. In this publication, Lanczos shows the philosophical meaning behind the great theories of Euler, Lagrange, Hamilton and Jacobi. In the Preface, he revealed that his foremost intention was "to give the student a chance to discover for himself the hidden beauty of these theories. . . ." This book, which Lanczos dedicated to Albert Einstein, was reprinted four times and became a standard text in universities.

During the McCarthy era of the early 1950s, the staff of the Institute of Numerical Analysis could not avoid security investigation. Lanczos himself had been accused of disloyalty, and was declared—for a while—*persona non grata*. He accepted the verdict in the spirit of

Éamon de Valera, John Wheeler and Lanczos, Dublin, 1972 (courtesy of Jasushi Takahashi).

Ecclesiastes, being no longer surprised by human bigotry and political persecution. He accepted Schrödinger's invitation to join the staff of the Dublin Institute for Advanced Studies (DIAS) as Senior Professor in the School of Theoretical Physics. This position was the very first permanent position in his life. He was 60. However, he still had 15 years of academic life ahead, because the official retirement age in the DIAS was 75.

The Institute suited Lanczos perfectly. The daily teatimes in the Library provided excellent occasions for discussions, and the absence of a formal lecturing requirement gave the opportunity to do full-time research. Lanczos returned to his 'first love' in science: the theory of relativity. Werner Israel, who was twice a visiting professor in the DIAS, remembered that Lanczos could be seen often even among colleagues, deep in thought.

Professor Synge*, who was the Director of the Institute at that time, had a special talent to engage Lanczos in conversation, even if Lanczos seemingly had no intention to do it. At one time, Synge began to talk to one of the colleagues in a voice loud enough to be overheard by Lanczos: "You know, if Whitehead's theory could predict at least the perihelion of the Mercury, I would prefer Whitehead's theory to Einstein's." Lanczos's reaction was like an explosion and he vehemently entered into a sharp and witty discussion for the pleasure of the scholars present.

Lanczos receiving an honorary doctorate from Trinity College, Dublin (courtesy of the *Irish Times*).

We should note that A N Whitehead (1861–1947) was an English philosopher. Although his intellectual importance lies mainly in philosophy itself, he did significant work in mathematics, theoretical physics and philosophy of science. He challenged the conceptual foundations of both the special and general theories of Einstein by offering an alternative rendering of the theory of relativity. Synge worked on Whitehead's theory around the early 1950s.

In the beginning of 1955, Lanczos almost left Dublin. He wrote a very frustrated letter to Einstein saying that the Director of the Institute, Schrödinger, was the most arrogant individual Lanczos had ever met. He added that Schrödinger was not interested in Lanczos's work at all, and he (Lanczos) was going back to the United States. Einstein never expressed his disagreement with Lanczos so strongly and never attempted to use his influence on him so much as in his reply. He said that Lanczos was going to commit the greatest foolishness in his life. The Institute and Ireland were for him. As a result, Lanczos accepted Einstein's advice. He remained in Ireland and found his home there.

He lived with his second wife, Ilse Hildebrand, a highly educated German woman, in an apartment of a fine Georgian house. In the evenings they often entertained groups of scholars and scientists from abroad, and discussions occurred on many topics, among them the arts, especially music. Professor Butler of the University of Manchester's Institute of Science and Technology recalled a conversation, quoted in Rodin's article, in which mathematicians and musicians were paired. This led him to ask Lanczos: With whom would *he* like to be identified? 'Schubert' was his immediate answer, 'for he had breadth, depth and romance.' The highlight of the evenings was always when Lanczos, who was a gifted pianist, sat at the piano and played.

It was during the Dublin period that Lanczos wrote seven of his eight books. He advocated the principle that if one reached an impasse, one should set aside the problem and do something else. Whenever he reached an impasse, Lanczos wrote a book. Among these seven books, there is a remarkable one entitled *Applied Analysis,* which became one of the most widely-used applied mathematical textbooks.

Lanczos's main field of interest remained the theory of relativity. The most important problem he attacked towards the end of his life was in connection with the geometry of the four-dimensional space–time world with the Pythagorean law, $\Delta s^2 = \Delta x^2 + \Delta y^2 + \Delta z^2 - c^2\Delta t^2$. The $+++-$ asymmetry of the signs troubled Lanczos very much. He had a very deep awareness of beauty in science. For him a 'beautiful' theory was in some sense more true than a 'not beautiful' theory. He could not understand that nature kept the symmetry of the signs in the correct way as $++$ and $+++$ in two and three dimensions and would do it in other ways in cosmic dimensions where beauty is most relevant. He must have felt himself as did Gauss who, when asked how his work was going, answered: "I know the results but I still do not know how I am going to get them."

In 1974 he had a glimmer of hope that he had a major breakthrough towards the goal to change the sign $-$ for the sign $+$ in nature's Pythagorean law. He was even more enthusiastic about further results, which he considered to be a solution of the problem that could give his idea definite support.

Around this time, he renewed contact with Hungary, his native land, and her scientists. On 16 June 1974, Lanczos arrived a second time to the country to pay a visit to the Rolad Eötvös University and to the University of Szeged. He arrived in a jubilant spirit and began lecturing extensively on the topic of nature's Pythagorean law on which his second paper had been accepted for publication in the journal *Foundations of Physics* **5**(1) (March 1975) 9–18. He looked frail but lively, full of plans for the future. Only on the very last day of his life did he show some indication that something was wrong. He observed with some regret: "I feel that I should be working harder". On the evening of 24 June 1974, in Budapest, he went to discuss the topic of a future TV interview when he became unwell. He was taken to a hospital nearby where he died of heart attack the early morning of 25 June. His funeral took place in Budapest on 5 July 1974.

On the centenary of Lanczos's birth in 1993, the College of Physical and Mathematical Sciences of North Carolina State University (NCSU), Raleigh, NC, in the United States where Lanczos was visiting professor three times, paid tribute his memory by organizing a Centenary Conference. The Conference took place at NCSU in December 1993. As a part of the celebration, NCSU published Lanczos's collected papers as *Cornelius Lanczos, Collected Published Papers with Commentaries* (General Editor: William R Davis) (College of Physical and Mathematical Sciences, North Carolina State University, Raleigh, NC, 1998).

Acknowledgment
The author appreciates the consultation in English with Professor Carmine Prioli, Associate Head, Department of English, North Carolina State University, Raleigh, NC, USA.

Further Reading

'In Memory of Cornelius Lanczos', Ervin Y Rodin (ed) *Computing and Mathematics with Applications* **1** (1975) 257–268. This was a memorial issue with biographical commemorations by colleagues, collaborators and students,

'Linear Systems in Self-Adjoint Form', Cornelius Lanczos, *The American Mathematical Monthly* **65** (1958) 665–679.

Lanczos's two most successful textbooks were *The Variational Principles of Mechanics*, Mathematical Expositions No. 4 (University of Toronto Press, 1949) and *Applied Analysis* (Prentice-Hall, Englewood Cliffs, NJ, 1956).

John Synge

1897–1995

"Αὐτός ἔφα"

Petros S Florides

John Lighton Synge was arguably the greatest Irish mathematician and theoretical physicist since Sir William Rowan Hamilton* (1805–1865). He was a prolific researcher of great originality and versatility, and a writer of great lucidity and striking 'clarity of expression'. He made outstanding contributions to a vast range of subjects and particularly to Einstein's theory of relativity. His approach to relativity, and to theoretical physics in general, is characterized by his extraordinary geometrical insight. In addition to bringing clarity and new insights to relativity, his geometrical approach influenced profoundly the development of the subject from the early 1960s.

Synge was born in Dublin on 23 March 1897, the youngest of a family of four, Ada Kathleen Frances, Edward Hutchinson and Victor Millington being the others. He was born with a growth on the cornea of his left eye which, as a result of surgery, became useless for reading thereafter. At that time the family lived in Kingscourt, County Cavan, where his father, Edward Synge (1859–1939), was land agent of a number of estates, including that of Lord Gormanston, a one-time Governor of Tasmania. The Synges were sufficiently well off to employ, in addition to the usual domestic servants and gardener, nurses, governesses and a live-in tutor for the early education of their children.

Synge's mother was Ellen Frances Price (1861–1935), daughter of the distinguished Irish engineer James Price. Synge maintained that, in so far as it was genetic, his interest in mathematics was inherited from this side of the family. The Price family can be traced back to the Stuarts of Scotland and, in particular, to Sir William Stuart who settled in Ireland in the early seventeenth century.

In the male line, Synge's family can be traced back to the fifteenth century (and, indeed, further) to Thomas Millington, 'corruptly called Singe of Bridgnorth' in Shropshire. Originally the family surname ranged over Synge, Syng, Singe and Sing but the present form Synge was well established by 1600. According to tradition, the changing of the name from Millington originated with Henry VIII, who commanded

J L Synge, taken from a group photograph of the Dublin University Mathematical Society 1927–1928. He was then a Fellow and Professor of Natural Philosophy.

a favourite choirboy to 'Singe, Millington, Singe'. In view of the ambivalence in the pronunciation of the name Synge, it is helpful to point out that it is always pronounced to rhyme with 'sing'.

Synge's family were members of the Church of Ireland, and great many of his distant ancestors attained high Office in the Church. Pre-eminent among them was his direct ancestor Edward Synge (1659–1741), Archbishop of Tuam. J L Synge's great-grandfather, John Synge (1788–1845), married Isabella Hamilton, a granddaughter of Hugh Hamilton (1729–1805) (no relation to Sir William Rowan Hamilton). Hugh Hamilton was undoubtedly the most distinguished distant ancestor of Synge. Although Synge had some reservations as to the greatness of Hugh Hamilton as a mathematician, he did write that: "he was the most intellectual Irish Bishop of the eighteenth century, a better mathematician than Berkeley, a better physicist, a better theologian, and in practical benevolence not behind." We shall dwell briefly on Hugh Hamilton's life in order to bring out the amazing similarity between his academic career and that of J L Synge, and the absolute contrast between the religious beliefs of the two men.

Hugh Hamilton entered Trinity College, Dublin, at the age of 14, where he was subsequently elected a Fellow of the College at 22 and Erasmus Smith Professor of Natural Philosophy at 30. He was elected Fellow of the Royal Society of London (FRS) in 1761, and Member of the Royal Irish Academy in 1785; he was, in fact, an original member of the Academy at its foundation. On the ecclesiastical side of his career, he was appointed Dean of Armagh in 1768, consecrated Bishop of Clonfert in 1796, and translated to Ossory in 1798, where he died in 1805.

He wrote extensively on mathematics, physics and chemistry, and theology. His greatest contribution to mathematics was the publication of his *De Sectionibus Conicis* in 1758, a book on conic sections written in the strictly Euclidean approach. Although not strikingly original, the lucidity with which the book was written won it considerable acclaim at the time. The great mathematician Euler described it as a perfect book. The same can be said about his book *Four Introductory Lectures on Natural Philosophy* which was written in 1762; it remained a standard textbook in Dublin and Cambridge for over 40 years. Although Hamilton's original contributions to mathematics and theoretical physics bear no comparison, either in quality and quantity, to those of his illustrious descendant, it is not unreasonable to suggest that in Hugh Hamilton we may have another candidate from whom Synge might have inherited his interest in mathematics.

Distinguished as some of Synge's distant ancestors were, his most distinguished relatives belong to the recent past and the present. First, and foremost, is his uncle John Millington Synge (1871–1909), the world renowned playwright and poetic dramatist who portrayed, so vividly, movingly and beautifully, the primitive life of the Aran Islands and the western seaboard with 'sophisticated craftsmanship'. The command of language and the sheer beauty of prose that J L Synge displayed in all his writings must surely have been inherited from his playwright uncle.

Then comes his distant cousin Richard Lawrence Millington Synge, FRS (1914–1994), who, with Archer John Martin, developed paper partition chromatography in 1941; this is a quick and inexpensive method of separating the components of complex chemical mixtures. They shared the 1952 Nobel prize for chemistry for their work.

Last, but not least, is J L Synge's daughter, Cathleen Synge Morawetz, who was born in 1923. She is an eminent applied mathematician, and has made many pioneering contributions to partial differential equations and wave propagation. She has the unique distinction of having been the first woman to hold the Directorship of the famous Courant Institute of New York. She was, also, the first woman to be elected to the Applied Mathematics Section of the (American) National

Academy of Sciences in 1990, and only the second woman to be elected President of the American Mathematical Society in 1995. She was awarded the National Medal of Science in 1998.

To return to J L Synge himself, in 1903 the family moved closer to Dublin for the formal education of their children, first, for two years, to Bray, County Wicklow, then a favourite seaside resort some 12 miles south of Dublin; then to Sandycove just outside Dublin for eight years; and finally to Dundrum, then a suburb of Dublin. It was while in Sandycove, in 1909, that Synge acquired his first bicycle, 'a most important event in my life', as he put it. His uncle, the playwright J M Synge, died that year, leaving his much-used bicycle, among many other things, to the family. Victor, J L Synge's brother, got the bicycle, and Synge got Victor's old and dilapidated one. Synge was to pursue his two hobbies of cycling and sailing passionately throughout his life. Later on in his life he took up painting with some considerable success. Well after his retirement he also tried his hands on the mandolin but without much success.

It was soon after moving to Dundrum that Synge began to question his religious beliefs. His sister and brothers were confirmed in the Church of Ireland, but his brothers lapsed soon after. At the advanced age of 89, in his short unpublished autobiography, he wrote: "This infected me to such an obvious extent that my parents never, to my recollection, suggested that I should be confirmed. I never was." In stark contrast to the illustrious religious careers of so many of his ancestors, and in particular of Hugh Hamilton, Synge stated that "I became, and remain, an atheist, with no belief whatsoever in a God or gods of any sort."

In his book *Kandelman's Krim* (1957) he states, in his characteristic style, "I am a Protestant to the marrow of my bones, holding the essence of Protestantism to consist, not in the recitation of this creed or that, but in the assertion of the right of the individual to hold his own views on all matters and express them as he thinks fit, with the prudential reservation that one does not preach vegetarianism (at least not too violently) in the lion's den."

Synge's formal education began soon after the family settled at Sandycove. He spent two years at a very small dame school (a school with a single female teacher) close by, and then three years at Tudor House School, a small preparatory school in Dalkey, less than a mile from Sandycove. It was at Tudor House that he first mastered the elements of geometry and algebra. He moved to St Andrew's College, then on the north side of St Stephen's Green in the heart of Dublin, in 1911, where he remained until 1915. It was there that his mathematical ability surfaced.

He won a number of medals for mathematics, and also a number of exhibitions in cash; some of the money he spent on one of the best bicycles available at the time. He was also a keen footballer, and served as the secretary of the college football team.

He entered Trinity College Dublin in 1915, and at the end of his first year he won a Foundation Scholarship in mathematics. This was an extraordinary achievement, for, in those days, the Foundation Scholarship examinations were normally taken at the end of the third year. Besides the great honour attached to these Scholarships, Scholars were entitled to free evening meals (Commons) and free rooms in College. Synge duly moved into rooms in College where, away from family restrictions, he enjoyed his newly-found freedom. He formed close friendships, particularly with his fellow students C H Rowe and T S Broderick, who were later to become Professors of mathematics in Trinity College.

He also met Elizabeth Eleanor Mabel Allen, a history student who shared his religious and political beliefs. They became engaged in 1917, while they were both undergraduates. Thinking, falsely as it turned out, that their parents would disapprove of marriage, they got married clandestinely in July 1918 in a registry office ('certainly not in a church'); his brother Victor who recently graduated from the medical school and his friend Broderick were their witnesses. They had a 'wedding feast of scrambled eggs' in Synge's rooms in college and then a cycling honeymoon in County Donegal.

Of the mathematics courses then taught in Trinity, he found mechanics, hydrodynamics, elasticity and differential geometry particularly attractive, and very much to his taste; not so algebraic geometry which was so well established in College by the great geometer George Salmon (1819–1904).

He studied particularly closely the monumental and demanding treatise, *Analytical Dynamics*, of E T Whittaker (1873–1956). He found an error in the section dealing with small oscillations, which he duly communicated to Whittaker. Whittaker very politely acknowledged this, and made the necessary correction in the subsequent editions of the book. Synge had thus 'gained a patron', as he put it, who was to promote his election to the Royal Society of London in 1943.

Synge graduated in October 1919 with a Double Senior Moderatorship in Mathematics and Experimental Physics. He was also awarded a Large Gold Medal which, for financial reasons, he sold shortly afterwards. Following his graduation he was given a temporary job teaching mathematics to ex-soldiers rehabilitated, and in January 1920 he was appointed a College lecturer in mathematics. In that same year Trinity announced that it would elect a new Fellow, not on the basis of an examination, as was the normal practice, but on a *thesis*. Synge used his considerable expertise on analytical dynamics to investigate *The Stability of* [what else?] *the Bicycle*. A thesis was duly submitted, but the Fellowship was given to his friend Rowe.

In the late summer of 1920 Synge was appointed Assistant Professor of Mathematics at the University of Toronto. He and his wife arrived in Toronto in November and they were to stay there until 1925. The only well-known mathematician in the department was J C Fields, FRS, who by that time was not actively involved in research.

An event, however, took place in Toronto during that time which turned out to be of great importance to Synge; this was the meeting of the American Mathematical Society at the end of 1921. For the first time Synge came in touch with some of the leading mathematicians of North America 'who actually published papers and books'. In particular, he established good relations with the University of Princeton mathematicians O Veblen and L P Eisenhart, who at that time were turning their attention to Einstein's theory of relativity. So was Synge, who adopted Minkowski's elegant geometrical and visual approach to the theory.

In 1925 Synge was informed that Trinity College Dublin would elect a new Fellow (by *examination* this time) and that the Erasmus Smith Professorship of Natural Philosophy would become vacant. Synge resigned his Assistant Professorship in Toronto at the end of the academic year 1924–1925, and, with his wife and two daughters Margaret and Cathleen, returned to Dublin, where he was duly elected Fellow and Professor of Natural Philosophy. A year later he was elected Member of the Royal Irish Academy. The striking similarity in Synge's career, thus far, with that of his distant ancestor Hugh Hamilton is now apparent; Synge followed almost exactly the same steps that Hamilton did in his academic career almost 200 years earlier.

In his first year in Trinity, 1925–1926, he felt privileged to have in his class two brilliant students, A J McConnell, who was to become the Professor of Natural Philosophy and Provost of Trinity College, and E T S Walton*, who was to become the Professor of Physics in Trinity, and shared the Nobel prize for physics with J D Cockroft in 1951.

In 1926 Synge published his important paper, 'On the Geometry of Dynamics', in the *Philosophical Transactions of the Royal Society of London*. In this paper, he regarded the configuration space of a dynamical system as a Riemannian manifold (of *positive definite* metric), and used the method of tensor calculus throughout. A by-product of this work was the derivation of the important equation of *geodesic deviation*, an equation which gives, dynamically, the relative acceleration of two neighbouring particles. At the very same time, the Italian geometer and relativist Tulio Levi-Civita published the derivation of the same formula for an *indefinite* metric (his derivation, therefore, being applicable to the theory of relativity).

The most important undertaking of Synge during this time was the editing, with A W Conway, FRS, Professor of Mathematical Physics at University College Dublin, of the first volume of the mathematical papers of Sir William Rowan Hamilton*. This volume contained papers on *geometrical optics* and was published by the Royal Irish Academy in 1931. As a preparation for this work Synge produced an annotated catalogue of Hamilton's manuscripts and about 200 notebooks, no mean achievement. The expertise in Hamilton's work that Synge gained in the editing of this volume had a profound influence on many of his subsequent researches.

Although in the first five years after his return to Trinity Synge produced a number of important papers, mainly on differential geometry, dynamics and relativity, he found Trinity too slow for his 'restless spirit'; so when in 1930 he received an invitation to head the newly established Department of Applied Mathematics in the University of Toronto, he resigned his Professorship and Fellowship in Trinity, and returned, with his family, which now included the recently born daughter Isabel, to Toronto. Apart from a visiting Professorship at Princeton University, a number of short visits to Brown University in Maryland, and a brief appointment as a ballistics mathematician in the US Air Force during the war, he was to remain in Toronto until 1943. One of his brightest students was Alfred E Schild, with whom he wrote *Tensor Calculus* in 1949, a book very much in use to this day.

He continued his prolific research in many different fields, including now hydrodynamics, elasticity and electromagnetic theory, and he wrote his first two books, *Geometrical Optics: an Introduction to Hamilton's Method* in 1937, and *Principles of Mechanics* (with B A Griffith) in 1942; the latter is still widely used as an undergraduate textbook.

An important theorem, known now as Synge's Theorem, which he published in 1936, has been acclaimed as 'one of the most beautiful results in global differential geometry of the twentieth century'. He applied the theory of elasticity to investigate the problem of 'traumatic occlusion' connected with the periodontal membrane of an ordinary tooth. The result was the publication of a long paper entitled 'The Tightness of the Teeth, Considered as a Problem Concerning the Equilibrium of a Thin Incompressible Elastic Medium', published in 1933, and several shorter ones afterwards. When Synge's friend Charles Rowe, now Professor of Mathematics in Trinity, heard the title of the long paper he remarked: "His dentures must be troubling him."

It may be added that it was during his involvement with the US Air Force that his interest in *antenna theory* originated; it was to lead to the 'Synge–Albert antenna theory', which was published, with G E Albert, in 1948. As a ballistics mathematician he studied the dynamics of the spinning shell extensively. Contrary to the long-held view that there were only five forces acting on the shell, he discovered a sixth force, which, though very small, was subsequently observed experimentally.

An event of particular importance to the mathematical community took place in Toronto in 1932. Early in that year the now ageing Professor J C Fields, whom we met earlier, fell seriously ill. He sent for Synge to explain that he wanted to endow international medals in mathematics, a subject which, it may be recalled, was omitted from the Nobel prizes. He left it entirely to Synge to make all the arrangements, and to launch the Fields Medals at the International Congress of Mathematicians, which was held in Zurich in 1932; Fields died shortly afterwards in 1932. These medals are regarded as prizes for mathematics equivalent to the Nobel prizes, and they are awarded at four-year intervals. So as to emphasize the widely held belief that 'mathematics is a young man's game', the recipients must be under the age of 40.

In 1943 Synge accepted an invitation from the Ohio State University, Columbus, USA, to head the department of mathematics. He was to stay there for three years. There, in collaboration of W Prager of Brown University, he developed the method of the *hypercircle* for the solution of *boundary value* problems. A book on this method, *The Hypercircle in Mathematical Physics*, was published in 1957, and the method formed a precursor of the *finite element method* which is widely used in numerical analysis today.

Three years later, in 1946, he accepted an invitation to build up and head the Mathematics Department of the Carnegie Institute of Technology (now the Carnegie-Mellon University) in the industrial city of Pittsburgh. His most brilliant students there were the internationally renowned mathematicians R Bott and J F Nash. On a visit to Dublin in 1985, Bott, then professor at Harvard University, recalled that Synge cycled wearing a nose mask, to protest against the polluted atmosphere of Pittsburgh. Nash received the 1994 Nobel Prize in Economics, and the recent award-winning film *A Beautiful Mind* is based on his life.

In 1948 Synge bade the final farewell to North America to return to his native Dublin, having accepted a Senior Professorship at the School of Theoretical Physics of the Dublin Institute for Advanced Studies (in the company of Erwin Schrödinger*, Walter Heitler*, and later Cornelius Lanczos*). After his retirement in 1972 he was appointed Professor Emeritus of the Institute.

During his Institute years he focused his attention mainly, but not entirely, on relativity. His many contributions to the theory were (and still are) not only outstanding, but also highly influential, drawing research students, collaborators and eminent visitors to the Institute from all over the world. It was during Synge's tenure that the Institute for Advanced Studies became one of the great centres in relativity theory. Up to the mid-1960s, and primarily under his influence, about 12% of the world's relativists passed physically through the Institute. He encouraged and helped generations of students, many of whom distinguished themselves in the field of relativity.

J L Synge giving a lecture in 1985.

We have already mentioned that Synge's approach to relativity and, indeed, to theoretical physics in general, was characterized by his extraordinary geometrical insight. He felt just as much at home in the four-dimensional space–time of relativity, as in ordinary three-dimensional Euclidean space. In an astonishing paper in the *Proceedings of the Royal Irish Academy* (**53A**(6), 1950) he was able, for the first time, to penetrate and explore in detail the region inside the Schwarzschild radius, which is now called the black hole. At a time when many relativists thought that it didn't even make sense to talk about this region, this work was very remarkable indeed.

The almost universal geometrical approach to the theory of relativity in the last 40 years or so is due primarily to Synge's influence, especially to his two epoch-making books, *Relativity, the Special Theory*, published in 1956, and *Relativity, the General Theory*, published in 1960.

In 1992 the first J L Synge Public Lecture was given in Trinity College by Professor Sir Hermann Bondi, with the present author as chairman. To one of my introductory remarks, mentioned above, that by the mid-1960s 12% of the world's relativists passed through the Institute, Bondi said: "Every one of the other 88% has been deeply influenced by his geometric vision and the clarity of his expression. Some of us, I may say, have at times been daunted by this clarity because it sets a standard that the rest of us can strive for but it's very hard to obtain." The outstanding relativist, Sir Roger Penrose, who was originally an algebraic geometer, and, through him, Stephen Hawking, decided to go seriously into the field of relativity after reading Synge's books on the subject. Characteristically, Professor Synge had this to say in 1972: "If you were to ask me what I have contributed to the theory of relativity, I believe that I could claim to have emphasized its geometrical aspect."

Synge published 11 books, including the three absolutely fascinating and delightful semi-popular books, *Science, Sense and Nonsense* (1951), *Kandelman's Krim* (1957), and *Talking about Relativity* (1970). He published over two hundred papers, the last one at the age of 92; it was, appropriately enough, on geometry, his life-long love. He also wrote a number of delightful book reviews in the *Irish Times* in the 1970s. Every single book and every single paper is a remarkable work of art. A complete list of Synge's published works can be found in the book *General Relativity* (Clarendon Press, Oxford), edited by his former student L O'Raifeartaigh*; it was published in Synge's honour in 1972 on the occasion of his 75th birthday.

His geometric insight and 'clarity of expression' permeate all his scientific and semi-popular writings, and all his superb lectures and seminars. He was indeed a superb lecturer, as everybody who was fortunate enough to attend his lectures readily testifies. Writing in *The Times Educational Supplement* in 1974 on *Success in Teaching Mathematics,* Professor Sir William Hunter McCrea*, another outstanding Irish mathematician and theoretical physicist, stated that: "The greatest living lecturer in mathematics lives in Dublin. Readers who know his identity will surely agree with this categorical claim, even if they are in the top flight themselves. And if this Professor X recognizes himself, let him take these remarks as a humble tribute. Every lecture he gives is the superb performance of a master—or ought I here to say maestro?". There is no doubt that McCrea was referring to Synge.

It may be added that the word *maestro* is in no way misplaced. Synge, with his goatee beard, had a striking resemblance to the famous English orchestral conductor Sir Thomas Beecham. Professor Werner Israel, FRS, another outstanding student of Synge, gave the second J L Synge Public Lecture in 1994. He told the story of how when Synge, on a short visit to London in 1957, was walking in the neighbourhood of

J L Synge (right) with his former students A J McConnell (centre), geometer and one time Provost of Trinity College Dublin, and E T S Walton at Synge's 90th birthday party.

the Royal Festival Hall, a passer-by politely raised his hand and said: 'Good evening Sir Thomas.'

Synge's motto in all his writings, especially in his semi-popular ones, was: "The mind is at its best when at play", as he put it. He uses his extraordinary imagination, the 'clarity of expression', and the sheer beauty of his prose, to set the mind of the reader 'at play', at the same time imparting knowledge to the mind effortlessly and almost unconsciously.

Professor Synge was the recipient of many honours in his long life. Besides the ones already cited (FRS and Member of the Royal Irish Academy), he was a Fellow of the Royal Society of Canada, an Honorary Fellow of Trinity College, and President of the Royal Irish Academy from 1961 to 1964; he was awarded Honorary Doctorates from the University of St Andrews (1966), the Queen's University, Belfast (1969) and the National University of Ireland (1970). In 1943 he was awarded the Tory Medal of the Royal Society of Canada and in 1972 the Boyle Medal of the Royal Dublin Society. The Royal Society of Canada has founded a Mathematics Prize in his name and so has the University of Toronto. In Trinity College Dublin, his *alma mater*, we founded, in 1992, the J L Synge Public Lecture, and the J L Synge Prize in Mathematics, given in alternate years.

The present author feels hugely fortunate to have been one of the 12% of relativists who passed through the Institute for Advanced Studies. I had the immense privilege of working closely with Professor Synge as a Research Scholar from 1960 to 1962, and on and off for another eight years or so. It is not inappropriate, I think, to bring this article to an end with just two personal reminiscences, which are repeated from the author's obituary of 'Professor John Lighton Synge, FRS', *Irish Mathematical Society Bulletin* **37** (1996), which may throw some light on the working of his great mind.

Shortly after his wife died after a prolonged illness in 1985, Synge moved into a nursing home in Dublin, and I was more than happy to visit him from time to time. His mind was lively and vivid almost to the very end of his life. He continued reading three or four books a week and thinking about mathematical problems. On one of my visits just a few months before his death on 30 March, 1995, exactly a week after his 98th birthday, I was evidently surprised to see him reading a huge medical book on the circulation of blood. Seeing my surprise he said: ''Oh, I have some problems with the circulation of blood in my legs and I decided to learn something about it.''

On another visit, towards the end of 1993, he told me that the problem which occupied his mind at the time was Fermat's last theorem. When I ventured to say that: ''The problem was solved last July'', he said: ''Oh, I know that, but I am thinking of the problem from a different angle, in terms of the zeroes of the Fermat function $x^t + y^t - z^t$. You can think of t as a parameter and (x, y, z) as a point in a three dimensional space or you can think of (x, y, z, t) as a point in a four dimensional space.'' I don't know how far this approach would have led him, but it clearly indicated that his 'geometrical vision' remained undiminished to the very end.

Professor Synge was a kind and generous man. He encouraged, helped and inspired several generations of students, who will always remember him with gratitude, fondness, admiration and the deepest respect. In old age Synge suggested that a significant part of his epitaph might read: ''He encouraged younger men''. Alas, there is no tomb to engrave this, Synge having donated his body to the Medical School of Trinity College for research. It is, however, deeply and permanently engraved in the heart and mind of each one of his students.

Further Reading

Talking about Relativity, J L Synge (North Holland, Amsterdam, 1970).
Kandelman's Krim: A Realistic Fantasy, J L Synge (Cape, London, 1957).
Geometrical Optics: An Introduction to Hamilton's Method, J L Synge (Cambridge University Press, 1937).

John Desmond Bernal

1901–1971

Roy Johnston

Bernal was born on a farm near Nenagh, County Tipperary, on 10 May 1901; he died on 15 September 1971 in London after a long illness, having suffered a first stroke in 1963 and another in 1968.

He married Eileen Sprague, whom he had met while in Cambridge, in 1922. There were two sons, Michael (b 1926) who lectured in physics in Imperial College, and is now retired, and Egan (b 1930) who became a farmer in Suffolk.

Being somewhat polygamous by nature, he had another son Martin (b 1937) by Margaret Gardiner, and a daughter Jane (b 1953) by Margot Heinemann. Martin Bernal has specialized in linguistics and ancient history, and is the author of *Black Athena*, a controversial book which links Greek civilization primarily to Africa; he is currently in Cornell University. Jane Bernal is a consultant psychiatrist at St George's Hospital Medical School, University of London.

Bernal lived until 1922 at the family farm, Brookwatson, on the Portumna road, near Nenagh. He subsequently lived in Cambridge and in London, at various addresses, currently under plaque consideration by English Heritage.

He became a Fellow of the Royal Society in 1937, and then Professor of Physics, Birkbeck College, London University, in 1938. He was Scientific Adviser to Combined Operations under Lord Mountbatten from 1943. Post-war he became President of the World Peace Council, an honorary post which he held from 1958 to 1963 in succession to Frederic Joliot-Curie.

The Bernal family has Sephardic Jewish origins; they are on record in Limerick from the 1840s, where his grandfather John Bernal was an auctioneer, furniture dealer, and railway company director. His father Samuel emigrated in Australia in 1884 and worked a sheep-farm, returning in 1898 to live with his elder sister, Mrs Riggs-Miller, near Nenagh, helping with the management of her farm. Shortly afterwards he bought the farm Brookwatson, also near Nenagh, where J D Bernal was born.

Bernal's mother Elizabeth Miller was American, with a California and New Orleans background, and experience of Stanford and the

John Desmond Bernal, 1932 (photo: Lettice Ramsey).

Sorbonne. Sam had met her in Belgium, when travelling the continent with his sister. Her parents were Presbyterian, her father being a Minister in San José, California. Elizabeth, according to the requirement of the time, converted to Catholicism to marry Samuel, much to her parents' disapproval. The Bernal family had become Catholic in their Limerick period. (Much of this information on Bernal's family background was researched by Ann Synge; for further details, see her chapter in the Swann–Aprahamian biography cited in Further Reading.)

Bernal was initially educated locally; he and his younger brother Kevin went first to the Nenagh convent school, then to the Protestant school in Barrack Street, this being regarded as preferable to the boys' school run by the Christian Brothers.

Bernal picked up locally an early interest in science. In his teens he was aware of the Birr telescope, with which the Earl of Rosse* had some 60 years previously pushed forward the frontiers of telescope design, and was in touch with several other local gentleman-amateur scientists, one of whom, Launcelot Bayly, introduced him to crystallography; with another, one Parker, Bernal went geologizing. He developed a feel for industrial technology through contacts with local industry and the mine works at Shalee.

His mother had wanted to support his scientific inclinations, and researched the Irish educational opportunities, in the end sending him to boarding-school in England, initially to Stonyhurst, then later to Bedford, whence in 1919 he went to Cambridge. There was family religious pressure to take him out of the Protestant school, and the level of teaching of science in Irish Catholic secondary schools, even in the elite Clongowes immortalized by James Joyce, was not up to the standard required by his mother; boarding-school in England was considered necessary.

He was, however, acutely aware of what was going on in Ireland, and observed it during his vacations, keeping a journal, which is archived in Cambridge. He recorded his support for Redmonite Home Rule, and subsequently for Sinn Fein, which position later under Cambridge influence evolved into Marxism and support for the Bolsheviks. The influence of his mother's Protestant background, and early exposure to interaction with the Nenagh Protestant community, helped him to avoid identifying the Irish national question with Catholicism, as many had done. During his Bedford period he was devoutly Catholic, but in Cambridge he recorded how he lost his faith sequentially: "first God, then Jesus, then the Virgin Mary, and lastly the rites ... now I had a quarrel with the Church because I could not help seeing it as an active agent of political reaction."

For some analysis of his evolution towards Marxism via Irish nationalism, also his early interests in science and technology in the Irish context, which helped to stimulate his interest in the 'science and society' domain, see the present author's contribution to the Swann–Aprahamian biography.

In Cambridge in his Natural Science Tripos, after a false start in mathematics from which he switched, he took Part 1 physics, chemistry, geology and mineralogy in 1922. He then went on to concentrate on crystallography, writing for his final-year undergraduate thesis a paper on 'The Vectorial Geometry of Space Lattices', and followed with 'The Analytic Theory of Point Systems', in which he developed a group-theoretic approach to the space lattices.

This work drew him to the attention of Sir William Bragg, who was then setting up his research team to take advantage of the X-ray diffraction techniques pioneered by von Laue. He worked with Bragg in the Royal Institution until 1927, contributing to the experimental technology by the design of the X-ray photogoniometer subsequently to be produced by Pye of Cambridge as the standard tool of the domain.

Bernal then went back to Cambridge in 1927 to a lectureship in structural crystallography, where for the next decade he worked on the structure of liquids, inventing the 'statistical geometric' approach to liquid modelling, and on solids of increasing complexity: pepsin, proteins, viruses, and identifying the type of helical structures which subsequently led to the discovery of DNA.

Politically, Bernal's student Marxism, picked up in the immediate aftermath of the Bolshevik Revolution in Russia, evolved into an increasingly positive attitude to science and the role of scientists as a political force. For more information on Bernal's political evolution, based on his extensive diaries, see the contribution of Fred Steward to the Swann–Aprahamian biography, and also Bernal's first book, *The World, the Flesh and the Devil*.

Bernal's work in crystallography with Bragg led to his appointment to the Chair of Physics at Birkbeck in 1938, succeeding P M S Blackett. He had little time to settle in, as the war started. In the lead-up to the war he had been from 1934 associated with the Cambridge Scientists Anti-War Group, which was instrumental in transforming the government's attitude to civil defence, analysing the effects of bombs on cities, using scientific modelling and later taking account of Spanish civil war experience. This led to an encounter with Sir John Anderson, the Minister responsible, at an Oxford conference, to which Bernal had been invited by Solly Zuckerman. Surprisingly Bernal and Anderson got on well, and this began Bernal's induction into the process of the application of scientific methods in the planning of the war, both at the strategic level, and in the analysis of tactics based on the existence of innovative weapon systems, such as radar. This innovative approach was later to become known as 'Operational Research' (OR).

In this capacity, as adviser to Lord Mountbatten, he was responsible for the analysis of the suitability of the Normandy beaches for landing troops and equipment, which he did using a combination of historical evidence, geological knowledge, aerial photographs and hydrodynamics of wave motion, resulting in the successful outcome. He also had a hand in the development of the Mulberry floating harbour.

During his Cambridge period he had, thanks to his political activities, picked up much experience of the interactions between science and government, with Marxist insight into the historical background. This led him to publish his seminal *Social Function of Science* in 1939, which was celebrated in the 1964 festschrift *The Science of Science*, edited by Goldsmith and McKay, as the founding text of the scientific study of science itself in a social context, which by then had begun to thrive.

After the war he returned to Birkbeck where he built up the crystallography group, to the extent that it merited a Chair, which he occupied from 1964 up to his retirement in 1968, though increasingly incapacitated. Eric Hobsbawm, the Marxist historian, attributes his first stroke in 1965 to the pressures of academic politics. During this period he supervised the work of Rosalind Franklin, who subsequently contributed to the DNA work for which Watson and Crick received the Nobel prize.

Bernal at work on the molecular structure of a liquid.

His declining years were spent subject to increasing communication difficulties, though he continued to publish scientific papers with collaborators up to 1969, his last paper being a letter to *Nature* (with Barnes, Cherry and Finney) on 'anomalous' water.

Politically after the war Bernal was increasingly isolated by the 'cold war' environment. He put much effort into the World Peace Council and to the nuclear disarmament movement. He was among the prime movers in initiating the Pugwash conference, which was an important communication channel between the USA and the USSR at leading science and government level during the worst period of the Cold War. In this process he kept in the background himself, not wishing to compromise Pugwash by his Marxist associations.

Bernal had a totally integrated and egalitarian approach to science and to politics; for him the work of the technician and craftsman was as important as that of the scientist. He regarded this egalitarian teamwork process as being his visionary model for the socialist society of the future, rather than the flawed state-centralist model in the east.

He participated in the Lysenko debates in the Engels Society, and the forum for Marxist scientists in Britain with J B S Haldane and

Khrushchev addressing the World Peace Council. Bernal is seated on his left.

Hyman Levy, mostly at the philosophical level, but finding it uncomfortable, turned his attention primarily to the peace movement and to the promotion of trade unionism among scientists, having been a founder member of the Association of Scientific Workers. Successive editions of his *Science in History* gave declining attention to Lysenko's significance.

Having burned his fingers with the Lysenko episode, in his review of Watson's *Double Helix* and of the period in 1968, Bernal concentrated on evaluating his own lab's relationship to the work, and how they had managed to 'miss the boat', despite having developed the key experimental technology. He was unable to make the leap into the ethical and political problems which subsequently have emerged on the fringes of molecular biology and its applications.

Due to Bernal's relative isolation during the Cold War, many of his ideas were developed as a sort of 'Bernalism without Bernal' in various 'science policy research units' during the later 1950s and 1960s. They were taken, with acclamation, to the US by Derek de Solla Price, a Bernal disciple. These units now flourish in Sussex, Edinburgh, Manchester and elsewhere, usually with some recognition of Bernal's influence. In the Irish context, a science policy unit exists in University College Dublin, in the foundation of which the late Professor Patrick Lynch had a hand. He was a co-author, along with the engineer H M S 'Dusty' Miller, of the 1964 Report of the Organization for Economic Cooperation and Development (OECD), *Science and Irish Economic Development*, which was consciously though implicitly Bernalist. To the present writer, both authors, on different occasions, explicitly admitted Bernal's influence in the OECD context. Publicly, though, due to his Marxism, Bernal had then somewhat the status of a 'non-person' in Ireland.

In Ireland, though, Bernal enjoyed something of a 'posthumous rehabilitation', in the form of a Royal Irish Academy discourse by his colleague Dorothy Hodgkin, FRS, which took place in 1980 (see below). The vote of thanks on that occasion was proposed by Tom Hardiman, then the Executive Chairman of the National Board for Science and Technology in Ireland.

Further Reading

'John Desmond Bernal', Dorothy Hodgkin, *Biographical Memoirs of Fellows of the Royal Society* **26** (December 1980). Hodgkin worked under Bernal in Cambridge in the early 1930s, and subsequently was awarded the Nobel prize for chemistry in 1964. See also her article on Bernal in *Proceedings of the Royal Irish Academy* **81B**(3) (September 1981).

Marxism and the Philosophy of Science, a Critical History, Helena Sheehan (Humanities Press International, 1985 and 1993). This book contains an interesting chapter on Bernal.

Sage, Maurice Goldsmith (Hutchinson, London, 1980). This biography of Bernal (whose nickname was 'Sage') is based largely on secondary sources, and did not have the support of the family.

J D Bernal: a Life in Science and Politics, Brenda Swann and Francis Aprahamian (Verso, London, 1999). This is a multi-author biography, with insights from people having first-hand experience of his multi-dimensional activity. Authors include Ritchie Calder, Eric Hobsbawm and Chris Freeman. As well as contributions mentioned in the text, Peter Trent, a colleague of Bernal from Birkbeck, provides an assessment of Bernal as a scientist. Hilary Rose of Bradford and City Universities, and her husband Steven Rose of the Open University, discuss Bernal's role in the establishment of research on science policy; more widely they throw very interesting light on the 'purple, red and blue' strands which Bernal had planned for his biography, representing the emotional, political and scientific aspects of his existence. Abridged versions of the first two chapters of the above, by Ann Synge and by the present writer, which were mentioned in the text, were published in *Notes and Records of the Royal Society* **46**(2) 267–278 and **47**(1) 93–101 respectively.

The Visible College, P G Werskey (Allan Lane, London 1978).

The Anglo Marxists, E A Roberts.

Bernal's own publications, apart from his numerous scientific papers, include:

The World, the Flesh and the Devil (Cape, 1929).
The Social Function of Science (Routledge and Kegan Paul, London, 1939).
The Freedom of Necessity (Routledge and Kegan Paul, London, 1949).
Marx and Science (Lawrence and Wishart, London, 1952).
Science and Industry in the 19th Century (Routledge and Kegan Paul, London, 1953; Indiana University Press 1970).
World Without War (Routledge and Kegan Paul, London, 1958).
Science in History (Watts, London, 3 volumes, 1954, 1957 and 1965).
The Origin of Life (Weidenfeld and Nicolson, London, 1967).
Also posthumously: *The Extension of Man: Physics Before 1900* (Weidenfeld and Nicolson, London, 1973).

Ernest Walton

1903–1995

Brian Cathcart

Ernest Thomas Sinton Walton, from Dungarvan in County Waterford, shared the 1951 Nobel prize for physics with Sir John Cockcroft for their breakthrough experiment in the artificial disintegration and transmutation of atomic nuclei. Working at the Cavendish Laboratory in Cambridge in 1932, they used high voltages to accelerate streams of protons, and with these they bombarded targets of the lighter elements—the first time that atomic nuclei were penetrated by means entirely under the experimenter's control. The effect of the bombardment was to change one element into another: lithium atoms absorbed protons and then split into pairs of helium atoms. They had thus realized the alchemists' dream.

More than that, the two men were able to measure the release of energy that occurred with this change, and thus provide the first experimental proof of the energy/mass equivalence described by Einstein. In the words of the Nobel citation, their work "introduced a totally new epoch in nuclear research", for it marked the beginning of the whole branch of science popularly known as atom-smashing. (See *Les Prix Nobel en 1951* in Further Reading.)

Walton was a product of Trinity College Dublin, where he had enjoyed a glittering undergraduate career, and after seven eventful years in Cambridge he returned there, in time becoming Erasmus Smith Professor of Natural and Experimental Philosophy, a post he held for nearly 30 years. When he died, on 25 June 1995, he was still the only Irishman to have won a Nobel prize in the sciences.

Though born in Dungarvan (6 October 1903), Walton spent his formative years in Ulster. His father, John Arthur Walton, was a Methodist minister originally from County Limerick, and his mother, who died when he was two, was Anna Elizabeth Sinton, daughter of a long-established County Armagh farming family. The Methodist clergy of those days moved frequently, and from 1909 onwards John Walton served parishes in the north.

In due course Ernest was sent as a boarder to Methodist College Belfast, where he proved an excellent student and won high marks for English, French, science and mathematics alike. It was towards the

Ernest Thomas Sinton Walton (courtesy of AIP Emilio Segrè Visual Archives, W F Meggers Gallery of Nobel Laureates).

latter subjects, however, that his nature and background inclined him. Walton senior had an amateur interest in science, particularly astronomy, so such matters were discussed at home, while the son was fascinated by mathematics—late in life he would recount how during a childhood illness he was given a geometry textbook to amuse him and became engrossed in the subject. He also loved carpentry and other forms of handiwork; tools were a favourite birthday present and over the years he built up a formidable workshop.

He entered Trinity in 1922 with the aid of an Armagh County Scholarship and a College 'Sizarship', or bursary, and carried all before him in mathematics and experimental science. Graduating with First Class Honours in both subjects in 1926, he stayed on to take an MSc in physics under John L Synge*, tackling a problem in hydrodynamics: the behaviour of the vortices produced when cylinders are drawn through water. Already a talented experimenter, Walton built an apparatus that yielded photographs of very high quality for the time, and subsequently published two studies in Irish journals. Before the Master's degree was awarded he applied for an 1851 Exhibition overseas student award, specifying that he wished to study atomic physics under Sir Ernest (later Lord) Rutherford at the Cavendish Laboratory. Though Walton had no relevant track record Rutherford eventually accepted him, impressed both by the enthusiasm of his referees in Dublin and by the skill shown in his work on vortices.

At that time the Cavendish led the world in experimental atomic physics. Rutherford, its Director, had probably done more than any other individual to open up the field, while alongside him worked F W Aston, the pioneer of mass spectroscopy, and C T R Wilson, inventor of the cloud chamber, as well as Sir James 'J J' Thomson, the discoverer of the electron. Thomson had preceded Rutherford as Director before becoming Master of Trinity College, Cambridge, but he continued to work in the laboratory almost every day. Also on the staff in 1927 were younger men destined for great distinction such as James Chadwick and Patrick Blackett. It was a powerhouse and Walton arrived at an auspicious moment.

Atomic physics, which began in the study of electricity, acquired true momentum with the discovery of radioactivity at the end of the nineteenth century. It was found, by Rutherford in particular, that particles naturally emitted by radioactive elements could, when aimed at targets and manipulated in other ways, reveal much about the working of the atom. A quarter-century of such experiments yielded a host of important discoveries but by 1927 it was apparent to some, most notably to Rutherford, that the returns were diminishing. If natural radiation can be thought of as cannon-fire, Rutherford was becoming frustrated at having to watch his cannons pick their own ammunition, set their own range and fire themselves. He wanted artillery that was under his control, and which he could use to dismantle atoms at will. And the target he had in mind was the nucleus, that complex of particles at the core of the atom.

On the theoretical front, meanwhile, the advent of quantum mechanics in the 1920s offered revolutionary new ways of understanding matter, especially at the scale of the molecule and below. Rutherford believed that vastly better experimental tools were needed to complement the new theoretical resource.

Into this vibrant, expectant world stepped Ernest Walton, and, after a short induction course, his first move was to propose an ingenious way of producing fast particles. Rutherford gave his blessing, and assigned Walton space in a room shared with two other researchers, one of whom was John Cockcroft. Six years older than Walton and with a background in industry, Cockcroft was already showing the executive abilities that marked his later career, so that his research work often took second place to his administrative activities. He remained, however, a very able scientist and, though their fields of study at this stage were quite different, he proved most helpful to the young Irishman.

Walton was attempting to accelerate electrons by spinning them in a circular electrical field, in the hope that a slingshot effect would yield fast projectiles useful in nuclear research. The project depended on his being able to produce and sustain that circular field, and contain the spinning electrons within it, but after a few months of work and a great deal of

Walton, Rutherford and Cockroft (UK Atomic Energy Authority, courtesy of AIP Emilio Segrè Visual Archives).

spectacular sparking he found himself stumped. He quickly turned to an alternative project, which involved accelerating electrons in straight lines through a succession of cylinders carrying high-frequency currents, but again his first attempts failed. Then, while he was developing improved apparatus, a new paper published by a Norwegian, Ralf Wideroe, showed his work had been overtaken.

To most scientists such setbacks would have been disheartening, but Walton appears to have been unconcerned. In fact neither of his ideas was fundamentally wrong, and both would be brought to success years later, the first as the betatron, and the second as the linear accelerator, so it could be said that he was ahead of his time. Most important was that Rutherford had formed a good opinion of him. Writing in early 1929 the Cavendish director declared the young Irishman 'an original and able man' who had 'tackled a very difficult problem with energy and skill'. (See *Nine Ulster Lives* in Further Reading.)

Developments were taking place that soon shifted Walton into a new line of research. In 1928 a Russian theoretician, George Gamow, was applying the new quantum mechanics to problems of the nucleus, and in particular to an apparent paradox of radioactivity: that particles of low energy were able to escape from the nucleus through or over barriers of higher energy. No one could explain this, although Rutherford had made an attempt based on classical principles. Gamow rejected Rutherford's view and looked at the problem instead as one of waves—quantum mechanics blurs the line between particles and waves (if it does not remove it entirely), thus permitting some things to happen which classical physics forbids. In this new framework it was possible for a low-energy entity to pass through a barrier of higher energy, though the chances might be extremely small, and in the case of natural radioactivity Gamow was able to estimate the probabilities.

The implications for nuclear research were considerable, since the barriers surrounding nuclei affect particles that are going in as well as those going out. It had hitherto been assumed that anyone wishing to force particles into the nucleus from outside would have to endow them with higher energies than those possessed by the protective barriers. These were measured in millions of volts—energies which, though just about attainable in a laboratory in 1928, could not yet be mastered sufficiently to be useful in nuclear experiments. If Gamow was right, however, and there was a measurable probability of particles 'tunnelling' through the barriers at lower energies, then that changed everything.

It was Cockcroft who spotted this and, after corresponding with Gamow, he calculated that it should be possible to compensate for the low probabilities by producing an intense stream of particles. In crude terms, if, say, you had one chance in a million of breaching the barrier at the lower energy levels, then if you fired a million particles at an atom, then one of them should get through. Cockcroft proposed building an apparatus to attempt this, and Rutherford, seeing a possible shortcut to the new research tool he had been dreaming of, gave his backing. Soon Walton abandoned his own project to help Cockcroft.

Their first scheme was to accelerate a stream of protons by using a direct current at up to 350 000 volts. Few of the components necessary for such an apparatus were available commercially, and many of the techniques were outside the Cavendish experience, so, besides buying a tailor-made generator from the firm of Metropolitan Vickers, the two men had to design and make almost everything themselves, often proceeding by trial and error.

It was soon clear that their chief difficulties lay with the rectifiers, which converted alternating current from the generator into direct current. For these the two men employed large glass tubes in the shape of rugby balls, similar to a kind then used in X-ray machines, but all too frequently the tubes broke, either in handling, under vacuum, or as a result of spark damage, and each time the whole affair had to be rebuilt from scratch. Slowly, however, they nursed their machine along and steadily higher voltages were achieved. By the spring of 1930 they were experimenting with proton sources, and by mid-year they produced their first proton beams at 250 000 volts. They even attempted a few

The Cockroft–Walton accelerator showing Walton seated to observe scintillations, April 1932 (courtesy of The Cavendish Laboratory, Cambridge).

experiments with lithium and beryllium, although without a clear idea of what they were looking for.

Then things began to go wrong: a transformer failure, more breakages of rectifier tubes, hitches with the proton source. By the end of 1930 both men were thinking of starting afresh on a larger machine with far higher tolerances, and a few months later fate intervened to give them their chance. They lost their room, which was transferred to another department, and in its place Rutherford gave Cockcroft and Walton a former lecture theatre where there was ample space to put their bigger ideas into practice. For this Cockcroft had designed an ingenious voltage-multiplier circuit which would give them up to 750 000 volts, while Walton developed much more robust rectifiers employing a 12-foot tower of thick glass cylinders instead of the horizontal array of rugby-ball tubes.

The new apparatus, which took shape by the end of 1931, comprised the generator, the rectifiers, a proton source along lines developed by Aston, an accelerator tube and, at the foot of the tube, housed in a wooden box lined with lead, the observation point. To these were attached a great deal of subsidiary equipment including vacuum pumps, spark gap spheres, and power sources. It is also right to mention the role played by 'Apiezon' oils and greases, newly developed by Metropolitan Vickers for use both in a new generation of vacuum pumps, and in Plasticene-type seals for high-vacuum apparatus. The company gave Cockcroft and Walton early samples, which proved invaluable.

By January 1932, though there were still problems, it was clear they had created a far more reliable and powerful machine. Long weeks were now spent in tests, calibration and improvements to the beam. There were also distractions: they wrote a paper for the Royal Society, and they made a patent application relevant to the power transmission industry. But while they were delighted with their apparatus, Rutherford was becoming impatient, and more than one Cavendish memoir of the time records him telling the two that he was fed up with their fiddling about and wanted to see some experimental results.

So it was that on 14 April Walton—working alone because Cockcroft was busy elsewhere—set up an experiment, placing a lithium target in the line of the beam, and beside it screens of zinc sulphide crystals. Ensuring that the various vacua were satisfactory, he then turned on the power and slowly brought it up to the desired level for this first attempt, which was not more than 250 000 volts. Next he crawled on hands and knees—a precaution to avoid electrocution—to the observation box and set his eye to the microscope and there, on the zinc sulphide screens, he immediately saw flashes of light characteristic of alpha particle impact. He summoned Cockcroft, who in turn summoned Rutherford. All three knew what this meant: protons with an atomic weight of 1 were entering the nuclei of lithium atoms with an atomic weight of 7 to create a new unit with a weight of 8. This unit was then splitting into two equal particles, each with a weight of 4—helium atoms, otherwise known as alpha particles.

Cockcroft and Walton were thus the first to split the atom in a controlled fashion, and by doing so they created a vastly superior means of probing the nucleus. It would be difficult to exaggerate the astonishment with which their success was greeted, and not only in the scientific world. The British newspaper that broke the story, *Reynolds News*, made it the front-page lead, something almost unknown for a scientific development, and in the week that followed the *New York Times*, to cite just one example, carried four lengthy articles on the experiment, declaring: "Never was a result more unexpected obtained." (*New York Times*, 8 May 1932). One irony is that the same results could have been achieved on their first apparatus as early as 1930.

If the scientists were amazed, the public response was different, for just as the two men opened the door to the nucleus, so they also opened the door to the debate about nuclear energy. Their atoms not only split in two but also released energy, and though far more energy had been required to cause the split, many commentators took this as a sign that nuclear power generation would in time become a reality. Fears also surfaced—"Let it be split," said the *Daily Mirror* on 4 May, "so long as it does not explode." Walton considered such notions misguided and it is fair to say that when nuclear weapons and energy arrived, they were less the fruit of atom-splitting than of research into the properties of uranium. Such thoughts, however, could not dim the public excitement at the time.

Walton, now a celebrity, remained at the crest of a scientific wave for a further two years, conducting experiments, engaging in academic controversy and, notably, attending a Solvay conference—a physics summit meeting—alongside the likes of Marie Curie, Niels Bohr and Werner Heisenberg. Then in 1934 he stepped out of the mainstream to return to Trinity College Dublin as a lecturer.

How far this withdrawal was deliberate is not clear. He had some hopes of continuing his research in Dublin, but he must also have known that neither the University nor the country could afford a cutting-edge programme, and in the event he never again published important nuclear work. His heart, however, had always been in Ireland, and he always expected to return to his *alma mater*, which now honoured him with a special fellowship. Perhaps most significantly the move and the new, steady job—no small matter in these Depression years—enabled him to marry Winifred Wilson, a kindergarten teacher in Waterford whom he had been courting from afar since 1930.

If Walton had left the mainstream, it had not quite abandoned him. In 1938 he and Cockcroft shared the Royal Society's Hughes Medal, and during the Second World War he was asked several times by old Cavendish friends to leave Ireland and take up war work. He declined these latter requests, not because of any objection in principle but because he could not be spared—Trinity was already acutely short-staffed and Walton himself was in demand for Irish government scientific work. He was not told what the war work would have involved, but in the first instance it was probably radar, and later it was certainly the atomic bomb—he was wanted for the Manhattan Project in the United States.

Then in 1951 came the award of the Nobel prize, which astonished, gratified and almost overwhelmed him—in a speech at the banquet that followed the presentation he declared it "an honour so great that, even yet, it is difficult for me to believe that it is true" (*Les Prix Nobel en 1951*). The delay since 1932, though long, was by no means unique in Nobel history, and may be explained by the emergence of atom-smashing as a discipline in its own right, particularly since the war. From the vantage point of 1951 there could be no doubt that this was of immense importance and was here to stay, and although Ernest Lawrence had already received the prize for the invention of the cyclotron, there could be no doubt that it was Cockcroft and Walton who took the first step in creating 'big physics'.

By this date the partnership between the two men may have appeared in a misleading light, so different were their subsequent careers. Sir John Cockcroft had become internationally known as the director of Britain's Atomic Research Establishment, while Walton worked in relative obscurity, but it would be wrong to underestimate the Irishman's contribution to the 1932 experiment. Cockcroft was an electrical engineer with valuable experience in industry, and he provided many of the important ideas, but he was not an especially good experimenter and because of other commitments he only worked on the apparatus part-time. The deft and ingenious Walton did the majority of the construction work, much of it beyond the existing state of the art, and it was no accident that on the crucial morning he was working alone in the laboratory.

That the partnership was a genuine and balanced one was acknowledged both by Cockcroft himself, and by Rutherford, whose opinion of the younger man was high. In a letter to the government's Department of Scientific and Industrial Research in 1930, the Cavendish Director had written of Walton as 'a man of exceptional ability', adding: "The work on which he is engaged with Dr Cockcroft should be developed as rapidly as possible, and Walton is the only student we have who has the ability and the technique to carry it out." (This interesting reference is from DSIR/2/386 at the Public Record Office, London.)

If Walton's work after his return to Dublin was not known on the wider stage, it was valued and appreciated in Ireland. As a teacher of physics he had rare gifts, and generations of students recall his remarkable ability simultaneously to demonstrate and explain complex experiments during lectures. He also oversaw a transformation of his department both in scale and quality, and contributed generously to university life and to a variety of government and other committees. Later, and particularly after his retirement in 1974, he was much in demand in the United States and elsewhere as a public speaker on the early history of particle physics, a role he came to enjoy.

Walton remained to the end, however, a modest, shy man of quiet habits, as well as a devout Methodist. His four children, brought up in the family home in Darty, South Dublin, all took up careers in science. The wider legacy that he and Cockcroft left to physics, meanwhile, may be seen in the hundreds of institutions and thousands of scientists engaged in particle accelerator research around the globe.

Further Reading

Cambridge Physics in the Thirties, John Hendry (ed) (Adam Hilger, Bristol, 1984).
'Ernest Walton, Atomic Scientist', Brian Cathcart: in *Nine Ulster Lives*, Gerard O'Brien and Peter Roebuck (eds) (Ulster Historical Association, Belfast, 1992).
Cockcroft and the Atom, Guy Hartcup and T E Allibone (Adam Hilger, Bristol, 1984).
Les Prix Nobel en 1951 (Stockholm 1952).

Walter Heitler

1904–1981

David Glass

Walter Heinrich Heitler was born in Karlsruhe on 2 January 1904. He spent the formative stages of his career studying physics at a time when the subject was in a state of flux due to the development of quantum mechanics. In those early years he studied and worked in places such as Munich, Berlin, Zurich and Göttingen, where many of the leading theoretical physicists of the day were to be found. Heitler's research demonstrated the power of the new mechanics by applying and developing it with great success to a number of diverse areas, including the theory of the chemical bond, radiation theory, cosmic rays and quantum field theory. The connection with Ireland is due to his time at the Dublin Institute for Advanced Study between 1941 and 1949 when he was a leading figure in the international physics community. After Dublin he returned to Zurich as professor of theoretical physics and in his later years his focus turned to philosophical issues relating to science and religion. He remained in Zurich until his death in 1981.

Heitler was the son of an engineering professor and developed an interest in science before the age of 12. In an interview with John Heilbron for the Archive for the History of Quantum Physics, he recalls some aspects of his childhood. At the age of about 12 he made a telescope, with which he could see the rings of Saturn, and later installed a chemical laboratory in the bathroom. At school the science teaching was poor and his physics teacher's main objective was to refute Einstein's theory of relativity. On one occasion, Heitler got into trouble with this teacher for reading a book by Einstein under his desk.

Much of his education was in Latin and Greek, and his introduction to Plato began a lifelong interest in philosophy. While still at school he attended a course of lectures by Professor Bredig in the Technische Hochschule in Karlsruhe, where he became acquainted with quantum theory. On finishing school he enrolled as a chemistry student at the Technische Hochschule, where he was particularly influenced by a mathematician called Professor Boehm, who taught him how to 'think exactly and to think properly'. He also had the opportunity to pursue his interest in philosophy, especially epistemology and philosophy of

Walter Heitler.

science, and notes that '[his] interest in science was always a desire for knowledge only, and it was never an interest in the practical side of science'.

Having become more interested in physics than chemistry, Heitler moved on to Berlin, where he would come in contact with Einstein, Planck and von Laue as well as other eminent scientists. He particularly enjoyed the lectures by Planck on thermodynamics, statistics and radiation theory, and by von Mises on applied mathematics. He also had the opportunity to talk with Einstein and became more aware of the difficulties in quantum theory. However, Heitler soon discovered that his previous education in Karlsruhe had not equipped him well enough, particularly in mathematics. Furthermore, he found the professors to be rather remote and realized that he would not get much help with a PhD.

As a result, he moved to Sommerfeld's department in Munich, and within two years he had completed a doctorate in concentrated chemical solutions with Herzfeld. Although this work was at the borderline between physics and chemistry, Heitler really wanted to work in theoretical physics, but this was delayed for two reasons. First, he did not have time to study the important papers of Schrödinger and

Heisenberg when they were published in 1926, since he was busy completing his PhD. Second, Sommerfeld obtained a Rockefeller Fellowship for him, but this was to work in physical chemistry with Bjerrum in Copenhagen. Less than six months after going to Copenhagen, he wrote to Schrödinger* in Zurich asking if he could take the rest of his fellowship there. Schrödinger agreed. In Zurich Heitler met Fritz London, which he described as a 'decisive turning point in my career'.

Heitler and London worked independently of Schrödinger, despite discussions in seminars and almost weekly visits to the countryside with him. However, they had become familiar with his papers on wave mechanics and decided to use this approach to calculate the van der Waals' forces between two atoms. They did this by considering the interaction between the charges of two hydrogen atoms and obtained the Coulomb integral; this, however, would not account for the van der Waals' forces, and, although attractive, it was not large enough to describe the homopolar bond. At this stage they had not included exchange, which Heisenberg had already discussed in the context of two electrons in a singe atom. They were puzzled by this for some time. Heitler describes their breakthrough:

"Then one day there was a very disagreeable day in Zurich;... the Föhn.... It's a very hot south wind, and it takes people in different ways.... I slept till very late in the morning, found I couldn't do any work at all, had a quick lunch, went to sleep again in the afternoon, and slept until five o'clock. When I woke up at five o'clock I had clearly—I still remember it as if it were yesterday—the picture before me of the two wave functions of two hydrogen molecules joined together with a plus and minus and with the exchange in it. So I was very excited, and I got up and thought it out."

Soon he called London and they started to work. By late in the night most of the paper had become clear, although they still had to work on the proper formulation of the Pauli exclusion principle. This paper, published in 1927 when Heitler was just 23 years old, showed that the homopolar bond could be explained in terms of quantum mechanics since the bond could only occur when the electrons had antiparallel spins.

The paper with London is the best known of Heitler's contributions to science, and had a huge impact in the study of chemistry due to its application of quantum mechanics. In his biography of London, listed in Further Reading, Gavroglu discusses the importance of this paper in the development of quantum chemistry and also notes reactions to the paper: Schrödinger was surprised since he had not expected his equation to describe the whole of chemistry; de Broglie and von Laue regarded the paper as a classic; and Pauling and Wilson described it as "the greatest single contribution to the clarification of the chemist's conception of valence".

Although this was the only joint paper that Heitler and London published, each of them continued to develop their ideas after leaving Zurich. Heitler began to use a group theoretical approach and in 1931, he wrote an important paper with Rumer that considered the valence structure of polyatomic molecules and showed how chemical formulae are related to quantum mechanical wave functions.

There was little correspondence between Heitler and London up until 1935, partly because of a disagreement over Heitler not having sought London's cooperation in some of the papers on group theory. However, by 1935 both of them were in England and they seriously considered returning to their work in quantum chemistry. London in particular felt that their work had not been fairly represented by a number of scientists including Mulliken and Pauling.

This episode in the history of quantum chemistry is discussed in some detail by Gavroglu who considers the conflicting approaches to theoretical chemistry: Heitler and London worked from first principles, while Pauling and Mulliken favoured semi-empirical methods. The correspondence between Heitler and London indicates the importance they attached to theoretical methodology, and their infuriation that their work was not given due credit in a number of articles. They were unhappy that in some cases their fundamental work was ignored, and in others it was criticized, while Mulliken's molecular orbital approach and the work of Slater and Pauling were praised. They considered writing a book on the subject to prevent certain scientists from 'falsifying history', but this idea was not developed and they moved on to different areas of work.

After his time in Zurich Heitler visited Sommerfeld in Munich to ask his advice about a job, although he also took time to explain some of the details of his recent work on the hydrogen bond. A position was arranged for Heitler as an assistant in Cologne, but on his way there he decided to stop in Göttingen to see Born who offered him an assistantship. Heitler described the atmosphere in Göttingen as 'simply delightful' and found that the interests of colleagues complemented his own work on molecules. He now had the opportunity to study group theory in more detail and also learned more about matrix mechanics, reading all the papers of Dirac in one night. There were also frequent visits to Bohr's institute in Copenhagen and it was at this stage Heitler became convinced of the Copenhagen interpretation of quantum mechanics.

As well as lecturing in a wide range of subjects and working on the chemical bond, Heitler wrote a paper with Herzberg in 1929 which came very close to predicting the existence of the neutron. However, by 1932 the main focus of work changed to quantum electrodynamics, the 'unsolved problem'. His idea was to work on high energy physics within the framework of quantum electrodynamics and Dirac's recent work on the positron. As a result he studied bremsstrahlung to see if theory broke down at high energies.

In 1933 Heitler, on account of his Jewish ancestry, had to leave Germany because Hitler had come to power. There had been a strong connection between Göttingen and Bristol, and so Born wrote a letter asking them to take some of the Göttingen academics, which resulted in Heitler obtaining a position in Bristol. Seven years later, after the fall of France to Hitler, a number of refugees in the Bristol physics department, including Heitler and his brother Hans, were interned on the Isle of Man for several months. By the end of 1940 all of them had been released. The University had applied for their release, partly on the grounds that it was important to continue work on nuclear physics given the possibility of constructing an atomic bomb. Around this time, however, Heitler and some of his colleagues declined the opportunity to work on the atomic bomb project.

Although Heitler was unhappy at having to leave Germany, new opportunities opened up in Bristol, since there were a number of theoreticians who were also interested in quantum electrodynamics and radiation theory. He was also able to keep track of the relevant experimental work carried out by Blackett and Powell, who had a group in Bristol working on cosmic rays.

One of Heitler's most important papers was published jointly with Bethe in 1934, and dealt with bremsstrahlung and pair production. This paper was extremely significant to the study of cosmic rays and the discovery of the muon. While the Bethe–Heitler theory worked well at low electron energies there was doubt as to whether it held for high energies. This was due to penetrating particles in cosmic rays, which seemed to contradict their theory.

Blackett, for example, believed the particles were electrons and that the Bethe-Heitler theory broke down for high energies. Alternatively, if one accepted the validity of the theory then the experimental data would provide evidence for the existence of a new particle. Eventually the experimental work of Anderson established the existence of such a particle with mass intermediate between that of an electron and a proton, which became known as the muon. The confirmation of the Bethe–Heitler theory at high energies formed a significant component of this work.

A closely related development led to another very important paper by Heitler, this time co-authored with Bhabha in 1937. Blackett discovered cosmic ray showers, which seemed to illustrate a further breakdown in theory. Heitler and Bhabha, however, explained the phenomenon in terms of quantum electrodynamics by their cascade theory, where an electron can lose energy through bremsstrahlung, leading to pair production which in turn leads to bremsstrahlung, and so on. Initially Blackett thought this was nonsense due to the existence of the penetrating particles as mentioned above.

However, around this time attention was drawn to a paper of Yukawa published in 1935, which had predicted a particle with intermediate mass. As noted earlier, the evidence for the new particle became stronger and at a lively meeting in Copenhagen everyone accepted that a new particle had been discovered. This included Blackett who at this stage also accepted the cascade theory. In 1938 Heitler published an important paper with Fröhlich and Kemmer on the theory of mesons, in which they were able to account for the anomalous magnetic moments of protons and neutrons. They were also able to predict the existence of a neutral meson, which was later discovered through its decay into two photons.

Later it became clear that the penetrating particle mentioned above was not the particle predicted by Yukawa. In 1947 Occhialini and Powell discovered pi-mesons (which are the Yukawa particles), and found that they decay into muons (the penetrating particles). Even here Heitler made a contribution since the experiment used a development of an experimental technique employed by Heitler some years earlier.

In 1941 Heitler took up a position as an Assistant Professor at the Dublin Institute for Advanced Study, where Schrödinger was the Director. The Institute had been set up in 1940 and the decision to offer Heitler a permanent position was made at the first meeting of the governing board of the School of Theoretical Physics. The first major seminar series at the Institute was held in July 1941 to coincide with Heitler's arrival and was a great success, bringing together physicists from all over Ireland. Shortly after coming to Ireland, Heitler married Kathleen Winifred Nicholson, who had worked in research in Bristol, and in 1946 their son Eric was born.

Heitler had a successful and happy time in Dublin and became an Irish citizen soon after arriving. During this period he was promoted to Professor in 1943 and succeeded Schrödinger as Director in 1946. He became a member of the Royal Irish Academy in 1943 and a Fellow of the Royal Society in 1948. He also played an important role in establishing the School of Cosmic Physics as part of the Institute in 1947. He remained in Dublin until 1949.

Heitler continued his work on quantum electrodynamics and cosmic rays while in Dublin, and during this time he had a considerable influence on Irish physics collaborating with both experimental and theoretical physicists. Although he extended his earlier contributions to the theory of mesons, his most significant piece of research in Dublin was his work on radiation damping. This theory enabled problematic divergences in quantum field theory to be avoided, and papers were published in collaboration with a number of physicists.

Paul Ewald, Max Born, Walter Heitler and Erwin Schrödinger in Dublin, 1943.

In his biography of Schrödinger, which is listed in Further Reading, Moore notes that Schrödinger wrote to Born telling him how pleased he was with Heitler, and the high regard with which he held him not only as a scientist, but 'as a man and a teacher'. Heitler's ability as a lecturer is also emphasized by Mott in his obituary of Heitler for the Royal Society (in Further Reading), where he notes that he is "remembered in Dublin for the clarity and interest of his lectures". He was also involved in modernizing the theoretical physics courses and gave lectures on wave mechanics for chemists. In fact, these lectures were published by Oxford University Press as a book entitled *Elementary Wave Mechanics*. Heitler had already established himself as an author of physics texts with his book, *The Quantum Theory of Radiation*, first published in 1936, which had become one of the main texts in quantum electrodynamics.

For a long time Heitler had been interested in philosophical issues and whilst in Dublin he contributed to a book on Einstein's thought, giving a very clear exposition of the Copenhagen interpretation of quantum mechanics. His emphasis on the role of the conscious observer illustrates that he had already given considerable thought to topics he would return to in later years. Heitler's wife confirmed to Mott that he was already thinking seriously about philosophical matters during his time in Dublin.

In 1949 Heitler was offered and accepted a Professorship in Theoretical Physics in Zurich, where he stayed for the rest of his life. Although he enjoyed his time in Dublin, it seems that he was keen to return to the German speaking world, and the offer of such a prestigious chair (previous incumbents of which included Einstein, von Laue, Debye and Schrödinger) was an ideal opportunity. On taking up the appointment he became Director of the Institute and retained this role until he retired at the age of 70.

He continued to work in quantum electrodynamics using the renormalization method to reconsider the problem of divergences. He successfully applied the theory to the natural breadth of spectral lines in collaboration with Arnous, did work on non-local interactions, and published a paper on detailed balancing in statistical mechanics. Heitler was dissatisfied with renormalization even though it had proved to be successful, and by the early 1960s his research in physics came to an end.

Heitler's interests increasingly turned to philosophy and the relationship between science and religion. He published a number of books and articles in this field, and it is very evident that he had thought deeply about these issues for many years. His best known book in this area was translated into a number of languages and in English is entitled *Man and Science*. In this book he discusses the nature of science and a number of themes come to the fore.

He argues that modern science has concentrated on causality to the exclusion of teleology, which is concerned with explanations in terms of design and purpose. This has led to a division between humans and science since we do not think and act causally, but rather in terms of aims. The use of causal determinism in the context of humans leads to a view in which we become like amoral machines, everything distinctively human having been excluded. He describes this inappropriate use of a causal approach as "one of the most dangerous tendencies of our time", and all the more so as it takes place under the guise of science.

He also makes the related point that modern science concentrates on the quantitative to the exclusion of the qualitative. After discussing the ideas of Goethe, who is better known for his literary works but also made contributions to science, Heitler argues that "present scientific methods are in principle limited". The exclusion of both teleological and qualitative concepts restricts science in such a way that it eliminates the human element. The main point to note about this restricted science is described by Heitler: "Thus a causal quantitative science is certainly not a complete picture, and hence is not a 'picture of the world'. It provides a partial picture, only an aspect of the world, a sort of *projection of the world on to a causal-quantitative plane.*"

It is clear that Heitler is not objecting to the notions of causality or the use of quantitative concepts, but the problem arises when this partial picture is mistaken for the 'world-picture'. In essence, he seems to be rejecting the idea that the methods of modern science are the only means of obtaining knowledge.

On the basis of the points raised above Heitler makes some rather radical attacks on the orthodoxy often presented by scientists. For example, he argues that progress in science will be restricted unless teleological considerations are introduced. In particular, while accepting the validity of evolution, he maintains that it cannot be accounted for on the basis of chance. As a first step to invoking teleology, he suggests the notion of an overall plan. As an example he argues that cell division is governed by the overall plan of the organ formed. Of course, teleological considerations soon give rise to metaphysical questions since, if there is design, it will be natural to ask, 'Who is the designer?' Heitler claims that there are also metaphysical implications in a causal approach (presumably because there must be a first cause). Since scientists often ignore these implications in the case of causality, he suggests that they could do the same in the case of teleology.

Another of Heitler's radical claims is concerned with the question of the moral neutrality of science. Everyone is aware of the ways in which scientific ideas can be used to bring about great suffering in the world. The usual response of scientists is that science is neither moral nor immoral, but amoral; the problem is not with science, but with how it is applied. Heitler, however, claims that science is also implicated in these activities. To qualify this, it is when science is claimed as a world-picture that it is implicated. To quote Heitler: "This science puts forward a claim to total validity, setting itself up to be the whole and only truth. But a partial truth that claims to be a whole truth may very well be immoral... this claim to total validity is in danger of destroying all reverence for human life."

No doubt Heitler had in mind the development and use of the atomic bomb, but many of his remarks about science and its relation to human life and ethical values seem prescient in the light of modern discussions about genetics. It is interesting to note that, while his writings on these topics were largely ignored by physicists, the medical community was much more interested in his ideas.

A final point to note concerns Heitler's view of the relationship between science and theology. He puts the conflict between the two down to unwarranted generalizations often made in each discipline. Regarding Heitler's own views, Mott notes that he became a member of the Swiss Reformed Church toward the end of his life, and that his beliefs "helped him in his terminal illness to await the future patiently and without complaint". He died in Zurich on 15 November 1981.

Walter Heitler received many honours during his lifetime. In addition to being a Member of the Royal Irish Academy and a Fellow of the Royal Society, he received the Max Planck medal in 1968, the Swiss Marcel Benoist prize in 1970, and the Gold Medal of the Humboldt Gesellschaft in 1979. He was awarded a number of honorary doctorates and also received a prize for his philosophical writings in 1977.

Acknowledgments
I would like to thank the Niels Bohr Library of the American Institute of Physics, College Park, MD, USA, for the loan of a transcript of John Heilbron's interview with Heitler.

Further Reading

(*All these works were used in the writing of this article.*)

Fritz London: A Scientific Biography, K Gavroglu (Cambridge University Press, 1995).

'The Americans, the Germans, and the Beginnings of Quantum Chemistry: The Confluence of Diverging Traditions', K Gavroglu and A Simoes, *Historical Studies in the Physical Sciences* **25** (1994) 1–63.

Archives for the History of Quantum Physics (American Institute of Physics), Interview of John Heilbron with Heitler, 1963.

Man and Science, W Heitler (Oliver and Boyd, Edinburgh and London, 1963).

Schrödinger: Life and Thought, W J Moore (Cambridge University Press, 1989).

'Walter Heinrich Heitler', N Mott, Biographical Memoirs of the Royal Society 141–151 (1982)

Walter Heitler: 1904–1981, L O'Raifeartaigh and G Rasche in: *Creators of Mathematics: The Irish Connection*, S K Houston (ed) (University of Dublin Press, 2000).

Albert Einstein: Philosopher-Scientist, P A Schilpp (ed) (New York, Tudor, 1949).

Early History of Cosmic Ray Studies, Y Sekido and Harry Elliot (Reidel, Dordrecht, Holland, 1985).

Sir William McCrea

1904–1999

Iwan Williams

I first met Bill McCrea on a cold windswept rail station in Bangor, North Wales, back in 1959. I was then an undergraduate and Secretary of the Student Mathematical Society, the President being Dr R A Newing. He had decided that the Mathematical Society needed reviving and that a good way doing this would be to invite a very well-known scientist to come and give a lecture. McCrea was suggested as this well-known scientist, who accepted my invitation. I learnt somewhat later that Dr Newing had in fact been one of the first two research students that Bill McCrea had supervised. This little incident illustrates the respect and fondness that all his research students had for McCrea. It also highlights the willingness of McCrea to help and encourage his ex-students, in this case by undertaking a long journey from Surrey, where he was Professor and Head of the Mathematics Department at Royal Holloway College, to Bangor.

The subject of his talk that day was *The Origin of the Solar System*, in which he outlined his slightly unorthodox view of the topic whereby both the sun and planets formed by the accumulation of a number of sub-condensations, or floccules as he called them. This theory very neatly solved the so-called 'angular momentum problem', the problem of explaining how such a small amount of angular momentum came to reside in the sun compared with the amount in the Planets.

In this theory, the sun accumulated from those sub-condensations that were heading, from random directions, more or less directly towards the location of the sun. The planets accumulated from those not heading towards the sun. Hence, the sun spins slowly and the planets possess the vast bulk of the angular momentum. It is interesting to note that such ideas are now being advanced again, particularly in relation to the formation of giant planets orbiting close to other stars.

As a consequence of being inspired by this talk and much encouragement from Newing, I became a research student of McCrea at Royal Holloway College, working on the origin of the planets (I had previously been heading for a research project in algebra!). A particular consequence of the ideas in this theory was that Jupiter-like gaseous planets were first formed, and heavier elements like iron silicates then settled to their

First meeting of the Governing Board of the School of Theoretical Physics, Dublin Institute for Advanced Studies (21 November 1940). From the left; Erwin Schrödinger, A J McConnell, Arthur Conway, D McGrianna, Éamon de Valera, William McCrea, Msgr Patrick Browne, Francis Hackett.

centre under the action of gravity. This was the first time that it had been recognized that solid grains could settle under gravity through gas, an idea that is now fundamental in the standard 'solar nebula' theory for planetary formation. This was, however, far from the only field that Bill McCrea worked in. Indeed he was the perfect counter-example to the saying: 'Jack of all trades, master of none' as I was soon to find out.

On arrival at Royal Holloway College, I met two students who had both recently obtained their doctorates under McCrea's supervision and were leaving the college. One was Petros Florides, who had completed his thesis on a topic in the area of relativity and cosmology and who was on the point of taking up a post at the Institute for Advanced Studies in Dublin (where he is still working). Some years previously McCrea had been instrumental in setting up this Advanced Study Institute and in bringing in Schrödinger to be its first Director. He also served on the governing body of this Institute from 1940 to 1950. See the book by Moore in Further Reading for more details regarding the early years of the Dublin Institute.

The other student leaving was Derek McNally, originally from Ireland, who had produced a thesis on various aspects of star formation in galaxies, and was on the point of moving to a Lectureship at University College London. He later became the General Secretary of the International Astronomical Union, and also both Secretary and Treasurer of the Royal Astronomical Society. When McNally took up the latter post, he managed 50% of the achievement of Bill McCrea, who, at various stages, held all four officer posts in that Society. He was President (1961–1963) and Foreign Corespondent (1968–1971), in addition to having been Secretary (1946–1949) and Treasurer (1976–1979). Sir Harold Spencer-Jones is the only other astronomer to have collected this 'full house' of RAS officer appointments. Of particular importance within the context of the Royal Astronomical Society was McCrea's contribution as Secretary in the immediate post-war years, when the foundations of the modern society were laid and the scientific activities of the Society began to flourish. (For further information, see the book by Taylor in Further Reading.)

During my time with McCrea at Royal Holloway College, there was a strong connection between the Mathematics Department and Ireland. Dr B Yates and Dr M R C McDowell were both on the lecturing staff of seven, and there was also a research student, from Ireland, Patrick Dolan, who was working on the General Theory of relativity, and who later became a lecturer at Imperial College. McCrea's record with research students was, by any standard, excellent. Virtually all students completed their theses on time, and most found permanent appointments within the University sector.

During my time as a research student alone, in addition to those already mentioned, McCrea had three other students: Agacy, who obtained a University post in Australia; Hilton, who went to University College Swansea; and Swaminarayan, who obtained a lecturing post at Queen Elizabeth College, London University. My 'replacement' as a research student was Michael Rowan-Robinson, now Professor at Imperial College – so much for a steady state where a person has to produce one replacement during their lifetime.

The secret of Bill McCrea's success as a supervisor of postgraduate students was that he made sure that he got to know the students as individuals, listened to their problems and difficulties, and, above all, treated everyone as equal human beings. He would never interrupt a conversation with research students in order to talk to somebody 'more important'. He encouraged people to build on their strengths, and was always willing to offer constructive suggestions regarding the way forward on a problem. However, he was also wise enough to know when to keep quiet and let the students work it out for themselves. In an age when it was fashionable to say: 'Here is an interesting problem, get on with it and come to see me in three years with your solution', he insisted on regular discussions, even if the student had no obvious difficulties.

He also had a very good memory, and could remember what any individual was working on and the results that they had obtained, without reference to any notes. This held for all his acquaintances, not just his current students; in gatherings, he could introduce a person whom he had not met for some time as: 'This is Dr X, he produced those interesting results on such and such'.

One oddity regarding Bill McCrea and his research students is that, despite many continuing their research in astronomy, he published very few joint papers with them once they had obtained their doctorates. This perhaps illustrates Bill McCrea's reluctance to take any credit for any piece of work unless he felt that he had fully contributed to the effort.

William Hunter McCrea was born in Dublin on 13 December 1904, son of Robert Hunter McCrea. His stay in Dublin was, however, short, the McCrea family moving to Kent in 1906, before settling in Chesterfield. He attended Chesterfield Grammar School, before entering Trinity College Cambridge in 1923, where he became a Wrangler, Rayleigh Prizeman, Sheepshank Exhibitioner and Isaac Newton Student.

He graduated in 1926, and started postgraduate studies under the supervision of R H Fowler, whereupon he immediately demonstrated, as already mentioned, his willingness to abandon the conventional viewpoint and follow original ideas. In the fifth century, Anaxagoras had suggested that the sun was mainly composed of iron, a suggestion that did not go down well with the Greeks; they banished him from Athens for putting forward such heretical ideas. However, this notion regarding the composition of the sun held sway until the twentieth century in a slightly modified form, for in the 1920s it was believed that the sun was essentially similar in composition to the earth (and indeed to all the other planets).

In the mid-1920s, strong hydrogen lines were discovered in the spectrum of the sun. In 1928, a German astronomer, Unsold, suggested that hydrogen was in fact the dominant component in the solar atmosphere. At that time, McCrea held the Rouse-Ball Travelling Fellow at Göttingen University in Germany. He demonstrated that there were a million times more molecules of hydrogen in the sun's atmosphere than molecules of all other substances put together, and was awarded his doctorate from Cambridge in 1929 for a thesis entitled *Problems Concerning the Outer Layers of the Sun.* This calculation, and the recognition that most stars were similarly composed of about 75% hydrogen, revolutionized both the study of stellar evolution and cosmology, eventually leading to the 'Big Bang Theory' for the origin of the universe.

On completing his postgraduate studies he was appointed Lecturer in Mathematics at Edinburgh University, where he met his wife to be, Marion Core. He married her in 1933, and they remained together for over 60 years until her death in 1995, bringing up a family of two daughters and a son. In 1932 he moved to London to take up the position of Reader and Assistant Professor of Mathematics at Imperial College, before being appointed as Professor at Queen's University Belfast in 1936. During this period, the bulk of his published work was in the area of relativistic cosmology, thus illustrating for the first time his versatility by moving well away from the field of his pre-doctoral studies.

Though never an observer himself, he placed great importance on observations, and two of his important papers from this period were on *Observable Relations in Relativistic Cosmology*. In these papers, he analysed what could be deduced about the structure of the universe from astronomical observations, without making *a priori* assumptions that the universe is isotropic and homogeneous.

Perhaps the most famous contribution of Bill McCrea to science came in this period with the publication in 1934 of the Milne–McCrea theorem. In this work, in collaboration with Edward A Milne, and published in the *Quarterly Journal of Mathematics* **5** 73, in 1934, it was demonstrated that the homogeneous and isotropic solutions of Einstein's field equations all had simple Newtonian analogues, that could be derived without recourse to the deep and technical mathematics required when using the relativistic route.

It is perhaps remarkable that it took so long for Newtonian cosmology to be developed, or perhaps that it required the unorthodox mind of Bill to take the first step! It should also be noted that Newtonian cosmology is still used today to tackle many problems relating to the universe. His monograph, *Relativity Physics*, published in this period, also followed the above logic of explaining relativity without the need of complicated mathematics. (See Further Reading.)

One of the fundamental notions of relativity is time dilation, that is, the idea that a moving clock goes slower than a stationary one. This gives rise to the so-called 'twin paradox'. One twin goes on a journey and then returns to his point of origin, while the other twin remains at this point throughout. Time dilation then suggests that the travelling twin should appear the younger of the two when they are re-united. However it may be argued that, if all motion is relative, the 'stationary' twin could be regarded as moving, and hence should also appear to be the younger.

Herbert Dingle regarded this paradox as proof of the incorrectness of relativity. McCrea, however, in a famous paper which was published in *Nature*, and is listed in Further Reading, maintained that one twin had accelerated while the other had not, thus making the argument invalid, since it is clear that the one that is accelerating is the one that should be regarded as moving. The exchanges between McCrea and Dingle regarding this paradox were many and heated, and have become part of the legend of the early days of relativity. Throughout, though, they remained friendly towards each other, which was a credit to both. (Much later, an experiment was conducted in which an atomic clock was flown around the world. McCrea was duly proved to have been correct.)

McCrea's stay at Queen's University Belfast was interrupted by the war, and he spent two years as Principal Experimental Officer with the Admiralty, including a period working on Operational Research with P M S Blackett's group. After the war, he moved to be Professor of Mathematics and Head of Department at Royal Holloway College, University of London, where in 1960 I became his research student. By that time, he had been elected a Fellow of the Royal Society in 1952, and awarded an Honorary DSc by the National University of Ireland in 1954. He had also published two further monographs, *Analytical Geometry of Three Dimensions* in 1942 and *Physics of the Sun and Stars* in 1950. This latter book became the backbone of many courses on stellar structure that are still taught today as part of degree courses at many universities.

McCrea remained at Royal Holloway College until 1966, when he moved to be a Research Professor at the newly created Astronomy Centre at the University of Sussex. For a number of years prior to this, no doubt inspired by the success of the Dublin Institute for Advanced Studies that he had help to form, McCrea had vigorously campaigned for a National Institute for Theoretical Astronomy. In this he failed, the powers that be instead deciding on setting up two centres, the Institute for Theoretical Astronomy in Cambridge, and the Astronomy Centre in Sussex. Despite the success of both, the latter under his guidance, McCrea maintained to the end that a National Institute should have been established.

While at Royal Holloway College, and later at the University of Sussex, he maintained contacts with universities in Ireland, mainly through being an external examiner, both for written examinations at undergraduate level and for thesis examinations at the postgraduate level. In recognition of this, he was awarded honorary Doctor of Science degrees from Queen's University Belfast in 1970 and from the University of Dublin in 1972. He was also awarded honorary doctorates at a number of other universities.

He formally retired from Sussex University in 1972, but in Bill's case, however, retirement was only a formality, and he continued to pursue his interest in a number of cosmological and astrophysical problems. He was a regular attendee at meetings of the Royal Astronomical Society, where speakers could generally expect him to ask an apparently simple question, but one that required a deep understanding of the subject in order to be answered successfully.

During this 'retirement' period he gained many further honours, for example receiving the gold medal of the Royal Astronomical Society in 1976, being made a Freeman of the City of London in 1988, and being

Bill McCrea in later years (drawn by Dr Jonathan Hare, The Creative Science Centre, Sussex University, 1996).

knighted on his 80th Birthday in 1994. During this period he also contributed to two historical works, *The Story of the Royal Greenwich Observatory*, published in 1975 and *A History of the Royal Astronomical Society*, published in 1987. He died on 25 April 1999.

So far, we have concentrated mainly on Bill as a distinguished scientist. This he undoubtedly was but he was much more. He was one of the most widely respected and best-loved astronomers in the world, a person who placed personal contact higher than scientific contact. He was as happy talking to new research students on a one-to-one basis as he was talking to the most distinguished dignitary. This warmth of spirit also made him an ideal bridge builder between the astronomers of the UK and Ireland following independence. The same skills came into play when he visited Argentina soon after the Falklands war. He liked meeting people and indeed this was his major interest. One of his great regrets was never having met with Albert Einstein.

He enjoyed belonging to scientific societies, and also to their informal dining clubs. He was for example the President of the Mathematical Association in 1973. He was a keen member of the Royal Astronomical Society Club for over 50 years and served as its President for ten years between 1968 and 1978. These were difficult years in which to be

President, for arguments over the admission of women as club members raged. Through the vision, tact and diplomacy of McCrea and others, Margaret Burbidge (Director of the Royal Greenwich Observatory) became the first woman member in 1973 without mass resignations of male members from the Club.

McCrea lived a long and full life. Throughout he had a certain presence about him that suggested he would go on forever, never apparently changing from one year to the next. He had outlived most of his contemporaries, but Bill did not dwell on this, for he simply made friends and admirers with the following generations of astronomers.

Further Reading

Schrödinger: Life and Thought, W J Moore (Cambridge University Press, 1989).

History of the Royal Astronomical Society, R J Taylor (Blackwells, Oxford, 1987),

Relativity Physics, W H McCrea (Methuen, London, 1935).

'The Clock Paradox in Relativity Theory', W H McCrea, *Nature* **167** (1951) 680.

Cormac Ó Ceallaigh

1912–1996

Mark McCartney

Cormac Ó Ceallaigh was born on 29 July 1912, the first child of the eminent Dublin obstetrician and early Irish historian, Séamas Ó Ceallaigh, and his wife Maire Cecilia. He studied at University College Dublin and at the Universities of Paris and Cambridge. He was the Senior Professor in the Cosmic Ray Section of the Dublin Institute for Advanced Studies for more than 30 years, and made significant contributions to experimental particle physics. He died in Dublin on 10 October 1996.

Cormac's university education was a whirlwind of success. He had been lazy at school, and was told by his father that he would forego a university education unless he worked over the summer for an Entrance Scholarship. This spurred him into action and he won an Entrance Scholarship to University College Dublin in 1930, graduating with First Class Honours in Physics and Chemistry in 1933. He was then awarded a College Postgraduate Scholarship, and in 1934 completed a research project on the balancing of valve circuits which led to the award of an MSc. Cormac's abilities as a physicist were clear, and on completion of his MSc he was awarded a University Travelling Studentship, which took him to Paris for a year to work under Pierre Auger studying cosmic radiation using Geiger counters and the Wilson cloud chamber.

In late 1935 Cormac moved to Cambridge to work under Lord Rutherford at the Cavendish Laboratory. His research centred on the disintegration of nuclei by slow neutrons, and the production of electron–positron pairs by high energy electrons. Once again Cormac's abilities were soon rewarded, and in 1936 Cambridge awarded him an 1851 Research Scholarship, which was an unusual award to someone who already had had a major scholarship. His success was announced to him by Rutherford in a chance encounter on the stairs with the words: "I see you've wangled it again O'Kelly!"

In 1937 Cormac returned to Ireland to take up a position as Lecturer in Experimental Physics at University College Cork. While in Cork he continued research using the Wilson cloud chamber, and was awarded a PhD in 1943, but this period of his working life was chiefly characterized by a heavy teaching load. His lectures were marked by their clarity, his

Cormac Ó Ceallaigh in youth and old age.

ability to pin-point the heart of the topic under discussion, a keen sense of humour (which could be irascibly caustic) and an encyclopaedic knowledge of languages and literature, all of which shone through, not only in the lecture theatre in Cork, but in later life as when he spoke at international conferences.

But in 1949, after 12 years at Cork, Cormac needed a change. Writing years later in the *Journal de Physique* he recalled; "In May of that year, having held a post in what was then a small college in Cork, notable mainly for the fact that it was the place where the great George Boole spent the last decade of his life and where he died in harness in 1864, I decided it was time to attempt to get back into the main stream of physics research."

He visited Imperial College London, the Clarendon Laboratory in Oxford, and the Atomic Energy Establishment at Harwell. He met some

Cormac Ó Ceallaigh launching balloons at Bristol.

old friends from Paris and the Cavendish, but there were no job offers forthcoming. Indeed one friend suggested that he was being 'slightly eccentric' to think that any research group would be interested in taking him on! Professor Cecil F Powell of Bristol University, however, was interested, and offered him a modest research post. Cormac took two years' leave of absence from Cork, and he and his wife Millie moved, at no small financial sacrifice, to Bristol and to what he later described as 'the great turning point in my life'.

The group at Bristol, led by Powell's infectious enthusiasm, was using nuclear emulsions to investigate the interactions of particles found in cosmic radiation. A nuclear emulsion is a specially produced photographic emulsion containing a high proportion of silver bromide. An ionizing particle passing through the emulsion leaves a track of silver grains which become visible after development. The emulsion plates were exposed to cosmic radiation by sending them aloft using hydrogen-filled balloons. The balloons, which could maintain flight at 90 000 feet for several hours, were frequently responsible for sightings of 'flying saucers'. When finally recovered, the emulsion plates were developed and examined.

The tracks produced by the passage, interaction and decay of fundamental particles through the emulsion could be observed and analysed under microscopes, and this was done by teams of young ladies, under the direction of Powell's wife Isobel. These teams were known as 'scanners' or more informally as 'Cecil's Beauty Chorus'. When Powell offered Cormac the job at Bristol, he also arranged for Cormac's wife Millie (who was also a science graduate) to be employed as a scanner.

Cormac was a thorough and painstakingly careful experimentalist, and it was these attitudes to his work which led directly to him making the first observation of the K meson in emulsion experiments, a particle event which Millie, as a scanner, had originally spotted as unusual. He examined over 700 particle decays to find just two K meson events. During this period Cormac also made important contributions to the understanding of the decay of π and μ mesons. During the 1950s these mesons formed part of a growing menagerie of particles, which were not finally ordered in terms of quarks and leptons until the 1960s. Today we understand the K and π mesons as being made up of quark-antiquark pairs, and the μ meson, which is now called a muon, is now known to be a lepton.

In 1951, his leave of absence over, Cormac returned to Cork. A new position of Research Lectureship in Nuclear Physics was created for him, a post which he held in conjunction with his Lectureship in Experimental Physics. He continued his research based on the nuclear photographic emulsions, and collaborated closely with the team at Bristol. He employed two scanners and borrowed equipment from the Dublin Institute for Advanced Studies (DIAS).

In 1953 the equipment was returned in person to Dublin, when Cormac was appointed Senior Professor in the School of Cosmic Physics at DIAS. Cormac created a research group, and suggested the establishment of a similar group at University College Dublin (UCD) which was duly formed. He continued close collaboration with Bristol and also began collaborating with other groups in London, Brussels, Berlin and Warsaw. Cormac, however, saw the link with Bristol as crucial to the work in UCD and DIAS. In a short history of the Cosmic Ray Section at DIAS, he states: "The viability of the two Dublin groups depended entirely on the relationship with Powell's Group at Bristol who provided, *gratis*, a share of the material which they had exposed in various ways. One of Powell's greatest achievements was the setting up of large international collaborations, a most efficient means of promoting rapid advances in the field of particle physics."

Two important results of the DIAS–Bristol collaboration were the first observation of the nuclear interaction of the Σ^- hyperon, and the confirmation of a K meson decay mode involving the creation of positrons. The main difference between this work and the work he had carried out in Bristol was that the source of the particles was not cosmic rays, but rather man-made particle accelerators.

During the academic year 1965–1966, Cormac was Visiting Professor at the Tata Institute of Fundamental Research in Bombay. Here he met P B Price who worked in the Research and Development Centre of the (American) General Electric Company. Price and his colleagues at GEC were collaborating with scientists at the Tata Institute regarding meteorites, and among other things were using certain plastics for detection of cosmic rays. Cormac became interested and a collaboration was forged between DIAS and Price and his colleagues in America.

Over the next three years Cormac and other members of staff from Cosmic Ray Section 'spent long periods in the uncongenial winter climate of upper New York State' where they established the suitability of a particular type of plastic (Lexan polycarbonate) as a detector. The advantages of plastic detectors were that they were cheap and light, and sheets of plastic could be assembled into blocks which could be made thick enough to bring some particles to rest. The plastic could be interleaved with photographic plates and degraders, and particle tracks could be recovered from the plastic by chemical etching.

Another joint project with Bristol began in 1970 with the plastic detectors being flown by balloon high into the stratosphere, this time to detect not elementary particles, but highly charged nuclei. Interest in the presence of highly charged nuclei in cosmic rays had been raised by the work of the Cambridge cosmologist, and friend of Cormac, Fred Hoyle, who, along with colleagues in 1957, suggested the mechanisms for how atomic nuclei are 'built' within stars. This process is called nucleosynthesis. The experimental observation of the presence and relative proportions of heavy nuclei was thus important to help verify the work of Hoyle and others.

In 1976 a project devised by members of the Cosmic Ray Section at DIAS, including Cormac, was selected by NASA to go aboard the 1984 space shuttle. The experiment apparatus involved over 12 square metres of plastic detector, weighing over one ton, again the aim being to search for the presence of highly charged nuclei in cosmic rays. The equipment was intended to stay in orbit for only one year, but due to various unforeseen circumstances (including the 1986 Challenger disaster) it remained for six years. A consequence of this was that the experiment provided by far the most prolific data available on heavy nuclei in cosmic rays. By the time that data had come back to earth, however, Cormac had retired from his senior professorship at DIAS.

Cormac was made a Member of the Royal Irish Academy in 1951, and in 1978 was awarded the Boyle Medal by the Royal Dublin Society. He served on the Council of the Royal Irish Academy, on the Council of the European Physical Society, and the Physics III Committee at CERN. He had many interests. He was a witty raconteur and a composer of humorous verses. A native Irish speaker, he was an excellent linguist and was fluent in many languages. He took a keen interest in meteorology and recorded annual patterns of rainfall in his garden.

Music was another of his passions, while his hobbies included cabinet making, at which he was highly skilled, and sailing, at which he was highly competitive. Indeed his wife and he raced in different classes of boat simply to avoid the stress of competing with each other. They sailed together around Ireland twice, and also cruised to the Western Isles of Scotland, where Cormac's ability to communicate in Irish with Scottish Gaelic speakers persuaded a Free Church mechanic to break the Sabbath and repair the boat's engine. As a young man in both Dublin and Cambridge he was a rower. In later life some of his happiest times were spent in County Donegal, where he gardened on the grand scale and greatly enjoyed the conversation of his neighbours.

His memory is honoured by the International Commission on Cosmic Rays with the biennial award of the Ó Ceallaigh medal for 'distinguished contributions to cosmic ray research'.

Acknowledgments
Thanks to Professor Alex Thompson and Ms Eimhear Clifton of DIAS for providing me with a wide range of materials relating to Cormac Ó Ceallaigh and his work at the Dublin Institute for Advanced Studies. Thanks also to Cormac's eldest daughter, Niamh, for providing photographs and details of her father's personal life.

Further Reading

Obituaries in *The (London) Independent*, 1 November 1996, and *The Irish Times*, 18 November 1996.

'A Contribution to the History of C.F. Powell's Group in the University of Bristol 1949–65', C Ó Ceallaigh, *Journal de Physique Colloque* **C8** (1982) 43, 185–189.

'The Cosmic Ray Section: A Brief History', C Ó Ceallaigh: in *Éamon de Valera Centenary* (Dublin Institute of Advanced Studies, 1982).

David Bates

1916–1994

Ray Flannery

David Robert Bates was born in Omagh, County Tyrone, Northern Ireland on 18 November 1916. With his older sister Margaret, he attended a one-room one-teacher school known as 'Miss Quiglie's'. At the age of nine, his mother, the former Mary Olive Shera of a farming family, brought him to Belfast, seeking better educational prospects. His father, Walter Vivian Bates, who originated from Mountrath, County Laois, Eire, remained in Omagh to continue his business as a pharmacist, but travelled twice a week to Belfast to be with his family.

David attended the Inchmarlo Preparatory School, and then the Royal Belfast Academical Institution (or Inst.), where he was rather unhappy, except for his last two years when the excellent mathematics and science departments fostered in him a love for science. His first love was chemistry, and he built a small chemistry laboratory in an outhouse at home, where he spent many contented hours.

He entered Queen's University in 1934 where: "I was enticed away from chemistry by the superb lectures given by George Emeleus on experimental physics and by Harrie Massey on Mathematical Physics." After earning his BSc degree with first class honours in both Experimental and Mathematical Physics in 1937, he was awarded his MSc degree in 1938 for a thesis entitled *Recombination in the Upper Atmosphere*, a subject which was to become one of his lifelong scientific interests.

Upon his appointment to a Chair of Applied Mathematics at University College London (UCL), Massey took his most promising graduate student, David, with him to enter the PhD programme. Because of the war, David never submitted a formal thesis but, by 1951, he had published so many important papers that UCL awarded him a DSc degree. In the same year, at the age of 35, UCL appointed him as a Reader in Physics, and shortly afterward he was appointed Professor and Head of Department of Applied Mathematics at Queen's University Belfast, his *alma mater*. He chose to remain at Queen's, and serve it with supreme loyalty and devotion, until the end of his days.

"A scientist who does not take his work as seriously as he can is likely to do little". Thus said David Robert Bates, who not only took his

DEPARTMENT OF PHYSICS, QUEEN'S UNIVERSITY OF BELFAST
(Teaching and Research Staff) 1938/1939

Front Row (left to right): S.F. Boys, J. Wylie, H.S.W. Massey, K.G. Emeleus (Head of Dept.), R.A. Buckingham, R.H. Sloane
Back Row (left to right): T. McFadden, D.R. Bates, E.B. Cathcart, J.W. Wilson, K.H. Harvey, L.S. Leech, J. Hamilton

science seriously, but also made monumental original contributions to atomic and molecular physics, to planetary and stellar atmospheres, and to astronomy. David Bates was a pioneering theoretical physicist and also a chemical physicist, and the architect of the internationally renowned School of Theoretical Atomic and Molecular Physics at Queen's University Belfast, where he was a distinguished Professor for 43 years.

Very quickly in his career he gained world recognition and international acclaim for his trailblazing research. His research laid the basic foundations and established the fundamental concepts required for modern understanding of atmospheric and ionospheric science, which today is termed the subject of 'aeronomy'. The evolution of aeronomy into the modern unified discipline of scientific inquiry and intellectual pursuit has been dominated by David Bates, who can now be aptly described as being 'the Father of Modern Aeronomy'.

With his students at Queen's University Belfast, David identified the great variety of atomic and molecular processes responsible for atmospheric behaviour, and performed calculations to understand fully the basic physics and chemistry of the proposed reaction mechanisms. To achieve this goal, he extended both quantum and classical theory to cover the range of physical interest, and developed new theory when none existed. So not only did atmospheric science benefit from his considerable talent and intellectual power, but his theoretical formulations of various scattering and reaction mechanisms have now become standard textbook knowledge of the subject.

His early collaboration with Sir Harrie Massey in the 1940s resulted in major papers which transformed the study of the earth's atmosphere into a quantitative discipline, capable of real scientific inquiry. They speculated on the processes occurring in the ionosphere, wrote them down, and then calculated their efficiency. In so doing, the basic theory of the mechanisms was developed, which is still valid and in operation today.

Bates also investigated the physics underlying the luminosity or brightness of the atmosphere. Based on his calculation that a single sodium atom would scatter the yellow photons emitted from the sun at the rate of one per second, Bates predicted in 1950 that a strong emission of yellow light would be seen if sodium were released at altitudes near 90 kilometres during twilight. He proposed the use of rockets to eject grenades which would produce localized sodium clouds, and estimated that an artificial cloud containing 1 kilogram of sodium vapour would produce a yellow cloud with illumination much greater than Venus at her most brilliant, and comparable with that of the full moon.

Bates' most celebrated experiment along these lines was conducted in 1955 and was certainly spectacular. The experiment was included as part of the UK space research programme and an impressive glow did indeed occur, much to the delight of the popular press, which ran headlines hailing the arrival of 'artificial moonlight'. Of these press accounts David, with his legendary dry wit and humour, reminisced as follows: "Nonsensical pieces were published on lighting up towns by sodium clouds. Photographers were sent to get pictures of me, reporters to interview me, or if that was a bother, to at least personalize the press release, which was easy because provincial professors could be assumed to be absent minded, untidy, and other-worldly. The folly stimulated *Pravda* to publish a skittish article entitled 'Moonshine Madness', according to the aged *Irish Times* correspondent in St Petersburg, a correspondent whose very name was a guarantee of the authenticity of the report—I refer of course to Rory O'Snowinhisbooze."

That experiment was an important stimulus to the UK space programme, but remained something of a bugbear for Bates, because of the celebrity status accorded him by the press. Although considered by now a visionary and luminary, he was a thoughtful retiring man who did not naturally invite such attention.

David and his wife Barbara, whom he married in 1956, enjoyed a great marriage, and two children, Katharine and Adam were born.

Fortunately for us then at Queen's (1958–66), David and Barbara extended their family, by including undergraduate and graduate students, postdoctoral fellows and visiting scientists from afar, who were bound together by their deep affection and respect for David and Barbara.

David was devoted to science and expected his graduate students and staff to share that commitment. It is no accident that the many graduate students and postdoctoral fellows in the Applied Mathematics Department at Queen's have continued to have a great impact on their subject as they progressed onwards to creative careers in many different countries. I found David to be one of the most humane people I have ever met, even to this day. He was such a warm, sympathetic, kind person, with a quick, engaging and wry sense of humour, but always ready to respond supportively to those in need.

I remember, on returning to Atlanta after an overseas trip, a letter from David awaited me which simply said 'Welcome home'. Also, on other occasions David would counsel me with 'Never post a letter you enjoyed writing'. That sage advice 'has saved me from myself on countless occasions', David recollected. David and Barbara, with their deep affection and respect for each other, set such a fine example which most of us attempt to emulate. They set standards, social and professional, that are unexcelled. They have influenced the lives of generations of students and visitors to Belfast over the past 50 years.

The impact of David's research during all of this time at Queen's University was profound. Within a few years of arriving from University College London in 1952, he had built an internationally acclaimed School. He remained modest about this tremendous accomplishment, attributing also to his good fortune in attracting from University College London to Belfast an incredibly gifted team, composed of Alex Dalgarno, Benjamin Moiseiwitsch and Alan Stewart, all of whom became scientists of great distinction.

He created a supportive yet challenging environment at Queen's, and many came from all over the world to carry out research in atomic, molecular and atmospheric physics, under his inspiration and direction. To have built this world-class department within such a short time was no mean achievement. It required great intellectual might and vision, incredible leadership and insight, sustained dedication and loyalty.

It did not go unnoticed that David encompassed all these qualities, and so after the first Sputnik was launched in 1958 by Russia, it came as no surprise that the various research agencies in the United States, the US Air Force Office of Scientific Research (AFOSR), the Office of Naval Research (ONR) and the Advanced Research Projects Agency (ARPA) should seek his guidance and counsel. He was, by then, established as simply the most dominant figure in aeronomy in the world.

David was well on his way towards delineating the earth's ionosphere, and the planetary atmospheres of Mars and Venus, in terms of the basic atomic and molecular scattering and reaction mechanisms. Also he knew how fast a perturbed ionosphere could return to equilibrium, and when radio transmission could be resumed. And he knew that the heat shielding of spacecraft upon re-entry into the atmosphere would ablate by friction, and send metallic atoms into the atmosphere, which in turn would affect the recombination of the charged constituents of the atmosphere.

A great advantage of the US funding of Bates' scientific programme in the 1960s was that the first digital electronic computer (called DEUCE) at Queen's could be acquired by his Department. Research was then tailored to maximize the capacity of DEUCE and this allowed the researchers at Queen's to pursue innovative research at the cutting edge of new physics.

David Bates' research career (1937–1994) spanned over half a century and covers half a dozen diverse areas where he made major contributions to our modern understanding of collision processes, and of planetary and stellar atmospheres. He tackled problems which were seemingly beyond everyone's reach. In his exploration of aeronomy and of atomic and molecular physics, David was like a great detective at work. He started with relentless pursuit of all the facts, and study of the original research papers. Some of these papers, upon careful examination, he had to dismiss, and this made news in itself. He employed highly original thinking in seeking all the possible solutions which fitted the facts of the difficult and complex puzzle on hand. He then composed the simplest mathematical model consistent with the physics. Finally, his formidable mathematical skill allowed him to drive through solutions of great utility, power and elegance. The scope of his scientific inquiry was immense, and his results have endured today. His papers serve as a powerful testament to the success of his magnificent and superb approach.

As has already been mentioned, David's first research problem was that of recombination in the upper atmosphere. During the daytime, the earth's atmosphere absorbs the ionizing (extreme ultra-violet) radiation from the sun and the earth's ionosphere (which extends upward from a height of about 50 km) is mainly formed by the process of photoionization of atomic oxygen O, the most abundant constituent of the atmosphere around 200 km.

$$h\nu + O \rightarrow O^{+} + e. \qquad (1)$$

Here the solar photons of energy $h\nu$ are sufficiently energetic to strip away an electron from O, producing a singly charged ion O^{+} and a free electron e. During the daytime, the concentration of electrons and ions build up, and would continue to do so unless the source of ionization, the sun, is 'switched off', as fortunately happens at night-time. Then some mechanism, called recombination, whereby the electrons re-attach themselves to the ions, must occur in order to maintain the observed balance of electron, ion and neutral atom number densities. When an electron e and a positively charged ion A^+ attempt to combine to form a neutral atom A, the electrons cannot just stick to the ion, because energy and momentum are not conserved by the reaction process $e + A^+ \rightarrow A$.

A possible mechanism is that recombination occurs via the emission of a photon. This process is represented by

$$e + A^+ \rightarrow A + h\nu \tag{2}$$

and is known as radiative recombination, since radiation provides the mechanism by which the excess energy is removed in order for recombination to occur. It was the commonly accepted process via which the photochemical balance between electrons, ions and neutrals was maintained. It is the process of photoionization (1) taken in reverse.

David's first research task in 1937 was to calculate the rates of (2), that is, to figure how fast radiative recombination occurred in the ionosphere. After over 1000 hours of calculation (using five-figure log tables), he concluded that the rates of radiative recombination were too small, by several orders of magnitude, to explain the much higher rates of recombination which were directly deduced from diurnal variations and from eclipse measurements.

At the same time, the atmospheric scientist Sidney Chapman believed that the observed geomagnetic variation required that the concentration of negative ions in the ionospheric E-region should be much higher than the concentration of electrons measured by the radio astronomers. Negative ions can be removed by *mutual neutralization*, $A^+ + B^- \rightarrow A + B$, and by three-body ion–ion recombination, $A^+ + B^- + M \rightarrow AB + M$.

For the removal of negative ions in the ionosphere, Bates and Massey considered only mutual neutralization, since there are not sufficient gas species M in the atmosphere to render the three-body ion–ion recombination process competitive.

The research on the negative-ion theory of the ionosphere was interrupted by World War Two; Bates went to the Admiralty Research Laboratory, Teddington, and worked with Massey, Francis Crick and other notables on mine detection devices and mine design. Aircraft-laid mines could be activated by the magnetic field of a passing ship. Bates investigated the possibility of using coils carrying electrical current suitably placed on ships so as to reduce their magnetic fields and thereby avoid destruction. Later Massey, then Deputy Chief Scientist of the Mine Design Department of HMS *Vernon* near Portsmouth, recruited David in

1941 to help with Britain's offensive mining effort. Relying on his mechanical engineering knowledge, he made notable improvements in the packing of mines to be laid by aircraft. Massey noted that David ''introduced a breath of fresh air into a moribund subject of great importance for the defence effort''.

After the war, Bates and Massey resumed their search for a solution of the high recombination rate problem of the ionosphere. By considering the production and loss mechanisms, Bates and Massey were able to show that negative ions were rare and that mutual neutralization was therefore unimportant. The negative ion theory of the ionosphere was so attractive that Massey later confessed, during celebrations marking his 70th birthday in 1978, that he would have dearly liked the theory to be true. But Bates' decisive work showed it to be untenable.

They examined and eliminated other possible recombination processes and postulated a new reaction sequence in 1947. Firstly, the atomic ions C^+ collided with neutral molecules AB to produce molecular ions AB^+ as represented by the process

$$C^+ + AB \rightarrow C + AB^+. \tag{3}$$

Here C^+ gets sufficiently close to AB that an electron bound within AB 'jumps ship', and becomes attached to C^+, forming a neutral C. The process (3) is called charge-transfer. The slow electrons then combine with the newly formed molecular ions, AB^+, via a process represented by

$$e + AB^+ \rightarrow A + B \tag{4}$$

which Bates called dissociative recombination.

The proposal that dissociative recombination is rapid was originally put forward rather hesitantly, because such processes were erroneously regarded at that time as occurring through the weak coupling of the nuclear and electronic motions of the combined collision system. It was not until during the course of an invited visit to Lyman Spitzer at Princeton University in 1950 that Bates envisaged how dissociative recombination works and he then derived a basic formula for its rate. Bates discovered that dissociative recombination was represented by the sequence

$$e + AB^+ \leftrightarrow (AB^{**})_r \rightarrow A + B^*. \tag{5}$$

Here the electron accelerating in the field of the ion excites a bound electron of the ion and is itself temporarily captured by the excited ion to form an intermediate complex $(AB^{**})_r$. This super-excited molecule is unstable—it can either revert to its original ionized state or fragment into two neutral particles. The additional electron–electron repulsion within $(AB^{**})_r$ can be sufficiently large to blow apart (dissociate) the

David Bates lecturing on 'Airglow and Auroral Atomic Processes' at Georgia Institute of Technology in 1981.

molecule into its two fragments, A and B, one of which is left in an electronically excited state B^*.

This process is very efficient and indeed very rapid, being several orders of magnitude faster than the radiative recombination (2) originally proposed. The fragmentation can only occur via quantal rearrangement of the electrons. A crude classical analogy might be when a fly (electron) lands on an elephant (molecular ion), the elephant immediately disintegrates, and its parts fly off in all directions! In classical mechanics, it cannot be so, but such is the power that can be released by quantal rearrangement of electrons within a super-excited molecule.

Bates was therefore able to unlock a basic secret of nature. He proved that ionization balance in the ionosphere was maintained by photoionization (1) during the day, and dissociative recombination (4) during the night. Also the green and red lines of the excited atomic oxygen O^* produced by dissociative recombination are visible in the nightglow. The same photo-chemical processes of photoionization and dissociative recombination were at work in the carbon dioxide and molecular nitrogen atmosphere of Mars, and the carbon dioxide atmosphere of Venus. Not only did Bates postulate the key process for electron loss in planetary atmospheres, but he also provided the theory of a basic process which finds application in many sub-areas of physics, such as plasma physics and gaseous discharges.

It was also during his short stay at Princeton that Bates uncovered another of nature's secrets: how molecules can be formed from the very low number densities of atomic species A and B in the interstellar medium. If there are no seed molecules and no dust, then David proposed that radiative association, represented by

$$A + B \rightarrow AB + h\nu, \tag{6}$$

could be the only mechanism. Here the photon carries away the excess energy which allows recombination to proceed. David formulated and developed the first accurate theory of radiative association, which he later extended in 1983 to cover the more difficult problem of radiative association of complex ions.

When there are enough atoms, molecules AB can also be formed by collision of A and B with third bodies M which drain away the energy which must be lost in order for A and B to combine, as in

$$A + B + M \rightarrow AB + M. \tag{7}$$

A classic problem with a long history had to do with how fast positively charged and negatively charged ions recombine as they move through a gas. Bates proposed a theory of the process, radically different from the theories originally advanced by Paul Langevin and Sir J J Thomson in 1903 and 1925, respectively. He reasoned that the ion pairs A^+B^- become deactivated by a sequence of collisions with the gas species M, not directly to the lowest bound state of AB, but indirectly via collisional connections through an intermediate block of high vibrationally-excited bound states AB^*. This theory advanced in the 1960s was effectively exact and is now standard for three-body ion–ion recombination.

Bates also made important contributions to plasma physics. A plasma contains electrons, ions and neutral atoms. When the charged particles recombine in a plasma, radiation is emitted and escapes from the plasma. This causes a non-equilibrium distribution of atoms $A(n)$ in excited states n to be produced by radiative recombination (2), and by three-body recombination $e + A^+ + e \rightarrow A(n) + e$. They were removed by collisional processes

$$\begin{aligned} e + A(n) &\rightarrow e + A(n') \\ e + A(n) &\rightarrow e + A^+ + e \end{aligned} \tag{8}$$

of excitation ($n' > n$), de-excitation ($n' < n$) and ionization, respectively. State n was also depleted by the spontaneous emission of a photon of energy $h\nu$ in the radiative decay process $A(n) \rightarrow A(n' < n) + h\nu$.

Bates (with Arthur Kingston and Peter McWhirter) advanced a revolutionary theory for the full rate of decay of the plasma, taking into account all of the underlying collisional and radiative mechanisms. This collisional–radiative theory met with widespread success, particularly in its application to astrophysical plasmas and to the thermonuclear fusion programme which seeks to provide new sources of energy. It has since kept many scientists busy in their laboratories throughout the world.

Mention must also be made of the Bates–Damgaard method for the calculation of the probabilities or oscillator strengths for radiative transitions $A(n) \rightarrow A(n') + h\nu$ between atomic levels n and n' of atom A. In response to the need of astrophysicists for information on solar and stellar spectra, David formulated a general method for the calculation of highly accurate oscillator strengths, a method that received widespread application. It is now part of the textbook literature. The work is among the top ten papers cited in physics.

In summary, David Bates discovered innovative mechanisms for recombination in ionized gases. He formulated and developed the theories of many basic recombination processes and performed calculations of their efficiencies over a wide range of physical conditions appropriate to many useful applications. These principles, now standard, are in much use today.

David made pioneering and classic contributions with enormous significance and far reaching implications for our future. His work drew attention, decades before it became fashionable in the 1970s, to the factors which influence the composition of the atmosphere, not only to short-lived photochemical products such as ozone but also to carbon monoxide, methane and nitrous oxide, gases which live much longer. As Michael McElroy so aptly said in the citation of David Bates' award of the 1987 Fleming Medal: "[David] recognized the importance and subtlety of the ties that link the atmosphere to life and to the world of soil and ocean and foresaw the emerging importance of industrial man as an influence on the global environment". A noteworthy goal for all students interested in advancing knowledge is to work with significant people on problems of significance. I found this epitomized my scientific interaction with David Bates.

David made a brief foray into the possibility of communication with extra-terrestrial intelligence (or CETI) after he had received, in his capacity as editor of the journal *Planetary and Space Science*, a paper submitted by Carl Sagan for publication. Although he began as a hopeful believer in the prospects for CETI, David's thoughtful, careful and balanced analysis of this issue left him with the inescapable conclusion that the chance of getting a reply to signals transmitted from the necessary dedicated power station on earth would be minute, unless the government, with its huge resources, were prepared to support the project for well over 1000 years, perhaps even 10 000 years. As David said wryly: "Government planning on such a protracted time scale is surely inconceivable." Nevertheless, his exploration of the feasibility of radio searches for extra-terrestrial civilizations was the basis for his many popular talks on the subject to lay audiences all over the world. His conclusion remains as yet unchallenged.

It should now come as no surprise to the reader that David was elected as a Fellow/Member of the major academies of the world: The Royal Irish Academy (1952), The Royal Society (1955), Honorary Foreign Member of the American Academy of Arts and Sciences (1974), Académie Royale de Belgique (1979), Foreign Member of the US National Academy of Science (1984), International Academy of Quantum Molecular Science (1985). He also received many honorary doctorate degrees from major universities.

The broad scope and diversity of his research was acclaimed on both sides of the Atlantic by his winning major awards: the Hughes Medal of the Royal Society (1970), the Chree Medal of the UK Institute of Physics (1978), the Gold Medal of the Royal Astronomical Society (1979) and the Fleming Medal of the American Geophysical Union (1987). He was knighted in 1978 for his services to science and education.

In 1992, the European Geophysical Society instituted the Sir David Bates Medal, which is presented to individuals who have made outstanding contributions to planetary and solar system sciences. In 2000, the Division of Atomic, Molecular, Optical and Plasma Physics of the UK Institute of Physics inaugurated the Sir David Bates Prize to be awarded to those scientists who through their research made significant advances in atomic, molecular, optical and plasma physics. These two awards are fitting commemoration and acknowledgement of David's pioneering work in those areas.

With Sir David's death on 5 January 1994, all—his university, his country, and world science—lost a pioneering leader of great distinction and excellence, a visionary who will be long remembered by all who were fortunate enough to meet him.

Further Reading

'Scientific Reminiscences', D R Bates, *International Journal of Quantum Chemistry: Quantum Chemistry Symposium* **17** (1983) 5.

Creators of Mathematics: The Irish Connection, S K Houston (ed) (University College Dublin Press, 2000) chapter 16.

'Sir David Bates, FRS', M J Seaton, *Quarterly Journal of the Royal Astronomical Society* **437** (1996) 181.

'Professor Sir David Bates MRIA FRS', P G Burke and D S F Crothers, *Comments in Atomic and Molecular Physics* **32** (1996) 127.

'Sir David Robert Bates', A Dalgarno, *Biographical Memoirs of Fellows of the Royal Society* **43** (1997) 47.

John Stewart Bell

1928–1990

Andrew Whitaker

John Stewart Bell was born in Belfast in 1928, and was brought up and educated in that city, but worked at Harwell and Malvern in England, and then from 1960 at CERN, the Centre for European Nuclear Research in Geneva, until his untimely death in 1990. He was employed to carry out research in accelerator and elementary particle physics, and produced excellent work, some of it of the highest importance, in these areas.

But he was also passionately concerned with the fundamental aspects of quantum theory, a topic which had lain dormant since the famous debate between Albert Einstein and Niels Bohr in the 1930s. In the 1960s, Bell took major steps in understanding the nature of the theory, and, by the end of the century, after 40 further years of conceptual and experimental analysis of his work, it was clear that he should be regarded as one of the greatest physicists of all time.

Bell was born in the Tate's Avenue area of the city, an area where the families of both his parents had lived for several generations. It was not an affluent area, most of the inhabitants, including Bell's forbears, engaging in such activities as small-scale trading, labouring and factory work. While it was clear that John Bell, his elder sister Ruby, and his younger brothers David and Robert were highly able, it was also clear that, in those days shortly before universal secondary education, it would be difficult for them to gain the qualifications for them to fulfil their promise.

Their father, also called John, was away for much of John Stewart's early teenage years with the British Army, but their mother, Annie, encouraged them to keep at their education. Even so, as events transpired, only John was able to stay at school beyond the age of 14. Nevertheless all his siblings had successful careers, David in particular qualifying as an electrical engineer by part-time study, and eventually becoming a lecturer at Lambton College in Canada, and the author of several successful textbooks.

From an early age John used encyclopaedias and other books of reference to generate large collections of facts on which he would lecture to any of his family prepared to listen; indeed he became known in the family as 'The Prof'. At the age of 11, he passed the qualifying examination for grammar school; his family could not afford for him to go to one of the more prestigious schools in Belfast, but enough money was found for him to attend the Belfast Technical High School. While the curriculum at this school was less academic and more practical than at grammar school, John enjoyed the more vocational courses, and achieved enough academic success that by the age of 16 he possessed the academic qualifications to enter Queen's University Belfast.

In fact, though, the minimum age to enter Queen's was 17. Bell put the extra year to benefit, spending the time as a technician in the physics teaching laboratory, where he greatly impressed the members of academic staff, Professor Karl Emeleus and Dr Robert Sloane. Both these men were devoted servants of Queen's; Emeleus, an able atomic physicist, came from Cambridge as a Lecturer in 1927, and was Professor from 1933 until 1967. From 1939, Sloane was his loyal assistant, Reader in Physics from 1948, and, like Emeleus, a Member of the Royal Irish Academy. They recognized Bell's talent, encouraged him to attend the first-year lectures while he was still a technician, and lent him books. The savings from this year helped him to pay his way through Queen's when he entered as a student in 1945.

Bell was an outstanding student, though his lecturers, at times, may have found him a handful. He himself has told how he clashed with Sloane on the meaning of Heisenberg's famous uncertainty principle. Mathematically this has the form $\Delta x \Delta p \approx h$, or, in words, the product of the uncertainties in position, x, and momentum, p, cannot be zero (as would be the case for Newtonian physics, where both quantities may be known and measured exactly and simultaneously), but must be at least of the order of Planck's constant, h.

But who or what, Bell persisted in asking, decides what either Δx or Δp actually is? It seems that Sloane had no answer, or, even worse, he tried various forms of words, adapted from the rather bland formulations of Niels Bohr, which Bell denounced as meaningless. Poor Dr Sloane—a man of great integrity, for whom moral rectitude was as important as good physics—was accused of dishonesty; he would not be the last to irritate Bell by 'woolly thinking' about quantum theory.

Bell obtained First Class Honours in experimental physics in 1948, and was able to stay on an extra year, achieving a second degree, again with First-Class Honours, in mathematical physics. His teacher here was the famous physicist Peter Paul Ewald, who had been one of the founders of X-ray crystallography 35 years before. He had been driven out of Germany by the Nazis, and spent the war years in Belfast, though soon after the war he moved to America, finishing his career at the New York Polytechnic University. Bell learned much from Ewald, and also enjoyed his irreverent sense of humour.

In 1949 Bell obtained a job in accelerator physics with the UK's Atomic Energy Research Establishment at Harwell, though he was to spend some time at an out-station at Malvern. This was an important time for elementary particle physics. Before the war the small number of elementary particles that were known had been discovered either in the laboratory (for example, the electron and neutron), or in cosmic rays (for example, the positron and muon). Cosmic rays were produced in other parts of the universe, and accelerated by high magnetic fields, arriving at earth with high energies. They had one great virtue—they were free! However their defect was the inconvenience in waiting for them to turn up (see the chapter on Ó Ceallaigh).

For this reason, even before the war Ernest Lawrence in the USA had developed the cyclotron, the first artificial accelerator, to provide, at will, high energy particles, which could impinge on targets liberating new elementary particles. Naturally such machines were expensive, but the convenience of having particles beams available permanently has meant that more and more powerful (and expensive) accelerators have been built over the past 50 years.

Bell's job was to model the paths of particles through accelerators, in particular studying instabilities in the beams. In those days before computers were available for such work, this required a sound knowledge of physical principles, particularly electromagnetism, and the ability to make suitable approximations so that problems might be solved without losing any of the essential physics. Bell's work was excellent, giving indications of a new focusing principle, strong focusing, which consisted of applying both axial and radial focusing. This maintained the integrity of the accelerated beam as energies were increased. When this technique became established, Bell rapidly became an expert, and he was a consultant to the team designing the proton synchrotron at CERN in 1952. CERN was being established because, as the cost of accelerators increased, no one West European state could afford to compete with the USA or the USSR; only a combined European initiative was feasible.

In 1953, Bell was offered leave for a year to work with Rudolf Peierls, another former German physicist who had had to flee Germany, and had become Professor of Physics at Birmingham University. Peierls was a great physicist, one of the very last who contributed in all fields of physics. He was to become a great friend and supporter of Bell. In Birmingham Bell worked on the *CPT* theorem in quantum field theory. Under the parity operator *P*, one imagines replacing a physical event by its reflection in a mirror; under the charge conjugation operator *C*, one replaces particles by anti-particles—electron by positron, proton by anti-proton and so on; and under the time reversal operator *T*, one may imagine a film of the event being shot and being shown backwards.

Until 1957 it was taken more or less for granted that the event obtained by any of these operations on a physical event would itself have been a proper physical event. Since then it has become clear that this is not so for *C*, *P* or *T* individually, nor for the operations *C* and *P* performed together. However the *CPT* theorem demonstrated that when all three of the operations are performed together, the fundamental laws of physics ensure that one does indeed obtain a proper physical event. This was an important result, and Bell produced a particularly clear proof, but unfortunately the very famous physicist Wolfgang Pauli, and also Gerhard Lüders, produced the result at roughly the same time, and they obtained the credit for what is always called the Lüders–Pauli theorem.

However, this work gained Bell a transfer to a new group at Harwell set up to study elementary particle physics, and also, after some further work, a PhD in 1956. Another important event at this time was his 1954 marriage to Mary Ross, a member of the accelerator design group. They were to have an extremely happy marriage, and wrote some papers together. By 1960, though, they became concerned that Harwell was moving away from pure research, and they both moved to CERN in Geneva.

Taking Harwell and CERN periods together, Bell published around 80 papers in the fields of elementary particle physics and quantum field theory. His most important work was carried out in the late 1960s with Roman Jackiw and dealt with an unsolved problem in quantum field theory. Theory appeared to predict that the neutral pion could not decay into two photons, but experimentally the decay took place. Bell and Jackiw, and independently Stephen Adler, were able to show that the usual approach using the standard current algebra model contained an ambiguity. Quantization led to a symmetry breaking of the model, the effect being referred to as the Bell–Adler–Jackiw anomaly.

Over the next 30 years, the study of such anomalies became important in very many areas of particle physics. It implied, for example, that each generation of particles must have a total spin of zero, and this in turn meant that there must be three 'colours' of quarks. In the first generation, for example, we have the electron (charge -1), the electronic neutrino (charge 0), and three colours each of up-quark (charge $\frac{2}{3}$) and down-quark (charge $-\frac{1}{3}$). Discovery of the anomaly eventually resulted in the award of the 1998 Dirac Medal of the International Centre for Theoretical Physics in Trieste to the surviving discoverers, Adler and Jackiw.

Another important contribution was Bell's 1967 suggestion that weak nuclear interactions should be described by a gauge theory, the kind of theory now used for all the fundamental interactions. The suggestion was taken up by Bell's collaborator, Martinus Veltman, whose research student, Gerard t' Hooft, showed that the idea worked in practice. (Unwanted infinities could be removed or 'renormalized'.) For this work, Veltman and t' Hooft shared the 1999 Nobel prize for physics.

While Bell was an important contributor to particle physics, it was quantum theory which made him famous. As we have seen, he thought deeply about the fundamental meaning of the theory even as an undergraduate. The first understanding of the need for a quantum theory had come from Max Planck in 1900, but it was not until 1925 that Erwin Schrödinger* and Werner Heisenberg produced the rigorous formalism for the theory, and by 1927 Bohr, with his complementarity approach, had sought to explain away the apparent conceptual difficulties of the theory. This was the Copenhagen interpretation of quantum theory, which satisfied nearly everybody except, of course, Einstein.

The most well-known problematic feature of the new quantum theory was its lack of determinism. The future behaviour of a system was probabilistic; in Einstein's famous words, God played dice. Yet there was a more deep-seated problem—lack of realism. It seemed that, except in special cases, a physical quantity did not have a particular value until it was measured; measurement actually created the value it observed. Indeed measuring devices appeared to have completely different properties from the atoms of which they were constructed, and the act of measurement seemed to be totally distinct from any other physical process.

Even as a student Bell realized that these problems could be avoided if one added extra variables, over and above those allowed for in the governing equation of quantum theory—the Schrödinger equation. These variables would—directly or indirectly—provide values for those physical quantities that lacked them. Such variables—*hidden variables* or *hidden parameters* as they are called—were anathema to the Copenhagen interpretation. More significantly, though, Bell read at this time of the famous mathematical argument of John von Neumann which claimed to prove that one could *not* add hidden variables to the structure of quantum theory. Von Neumann's book was written in 1932 but it was in German, which Bell could not read, and the book was not translated to English until 1955.

A further dimension to the problem had been provided by the Einstein–Podolsky–Rosen or EPR argument of 1935. In a convenient version of this argument suggested later by David Bohm, a particle without spin at the origin decays into two particles of spin-$\frac{1}{2}$ moving in opposite directions. It is a property of spin-$\frac{1}{2}$ particles, such as electrons or protons, that if you measure the component of spin along *any* direction, the answer must always be either $+\frac{1}{2}$ or $-\frac{1}{2}$.

Before any measurement, the wave-function of the system, which, according to Copenhagen contains *all* information about the combined system, tells us that we do not know what value we will obtain in a measurement of the z-component of spin for either particle; there is 50% probability of obtaining either of the two permissible values. However,

Trieste Conference in Honour of P A M Dirac's 70th birthday in 1972. In the foreground are Dirac and Werner Heisenberg. John Bell is sitting behind them on the extreme left (photograph by Foto 'Rice').

there is more to it than that: the wave-function is *entangled*, which means that if we measure $+\frac{1}{2}$ for the first particle, an immediate measurement for the second must give $-\frac{1}{2}$, and vice versa.

There seem to be two possibilities. The first is that there are indeed hidden variables, so that all possible measurement results for both particles are actually coded into the system right from the start. Otherwise it seems we must assume the information on the result of the measurement on the first particle is *immediately* available at the second particle to determine the result obtained there. The latter possibility implied propagation of the information of a measurement result at infinite speed, apparently in conflict with the assumptions of relativity, and appalled both Einstein and, later, Bell,

Bell's positive feelings towards hidden variables were increased yet further in 1952 when Bohm produced an explicit hidden variable theory which duplicated the results of quantum theory. His papers were derided by the physics establishment, but entranced Bell, who, as he later put it, "saw the impossible done". (His theory was also, incidentally, deterministic, though this was never a major issue for Bell.) Though Bell thought fairly consistently about quantum theory during this decade, it was not until 1963, when on leave from CERN at SLAC, the Stanford

Linear Accelerator Center in California, that he had the time to come to definitive conclusions and attempt to publish his work. He produced two masterpieces.

The first was a clear demonstration that von Neumann was wrong. Bell set up his own hidden variable model for measurement of any component of spin of a spin-$\frac{1}{2}$ particle; the model was simpler than that of Bohm, and too simple to be disregarded. He then proceeded to show exactly where the fault was in von Neumann's argument, which had survived so long. Von Neumann had, in fact, included a quite unnecessary axiom in his argument, without which it collapsed.

While Bell thus showed that hidden variables *were* possible, he also discussed two drawbacks to such theories. First, except in the simplest cases, he was able to show that hidden variables had to be *contextual*; the result obtained by measuring one observable depends on which other variables you choose to measure simultaneously. Second, he showed that, at least in Bohm's theory and several other hidden variable theories which Bell concocted, the physics was non-local: measurement results at one point depended instantaneously on events at all other points. He suggested that this unpleasant feature might be characteristic of all hidden variable theories.

This first great paper was written in 1964, but, as a result of a series of accidents, not published until 1966. In the meantime, Bell had published, in 1964 itself, his second great paper, which answered the question raised in the first in the affirmative: hidden variable theories of quantum theory *must* be non-local. His analysis was based on the EPR scheme described above, but generalized it so that the measurements in the two wings of the apparatus were of *different* components of spin. He was able to show that no hidden variable model of the situation was able to duplicate the experimental results predicted by quantum theory for *all* choices of measurements, unless one allowed the result in one wing to depend on the experimental setting in the other, but this would bring in non-locality.

This may be regarded as one of the great scientific discoveries of all time. The model of the universe that had lasted from the earliest days of science, with individual systems with well-defined properties interacting locally, had been shown to be wrong, at least assuming quantum theory was correct. By their refusal to contemplate the existence of hidden variables, both Bohr and von Neumann had accepted that the model had to be abandoned; Bell, though, had shown that their arguments were false, and provided his own correct alternative. Henry Stapp of the Lawrence National Berkeley Laboratory in California called Bell's work "the most profound discovery of science".

Perhaps, though, quantum theory was not correct, at least in the specialized EPR-type situations being considered. This was a highly

John Bell at CERN 1989 (photograph taken by Anita Corbin and John O'Grady).

interesting if perhaps unlikely possibility, which has been followed through by further theory and also much experiment. First John Clauser, Richard Holt, Abner Shimony and Michael Horne produced the so-called Bell–CHSH inequality for EPR experiments; it should be obeyed by theories with local realism, the word 'realism' merely meaning here the presence of hidden variables. The correctness or otherwise of quantum theory is not mentioned in this argument.

Three generations of experiments using pairs of photons with polarizations entangled in an EPR sense, associated principally with the names of Clauser and Ed Fry in the 1970s, Alain Aspect in the 1980s, and Anton Zeilinger in the 1990s have indicated that the inequality is violated, though there is still a remaining problem, since the efficiency of the detectors is low, and it is conceivable that they do not sample the photons fairly. It is hoped that tests with atoms or fairly massive elementary particles from accelerators will settle the matter fairly soon.

Bell's work had transformed the study of the foundations of quantum theory from one of 'armchair philosophy' to one of 'experimental philosophy'. By the end of the twentieth century, theory and experiment were working hand-in-hand to provide new perspectives and new practical possibilities; the area had become one of the most vibrant in physics. In addition it had given birth to the entirely new field of quantum information theory, including quantum computation, quantum cryptography and quantum teleportation, which will itself become a major field of study, both theoretically and technologically, for the twenty-first century.

Bell became a Fellow of the Royal Society in 1972, but he had to wait until near the end of his life for the majority of his honours. He was awarded the Dirac Medal of the Institute of Physics in 1988, and the Heineman Prize of the American Physical Society and the Hughes Medal of the Royal Society in 1989. In 1988 he received honorary degrees from both Queen's University Belfast and Trinity College Dublin. Had he lived ten years longer, he would almost certainly have been awarded the Nobel prize for physics.

Bell was always keen to help and, on several occasions, work with younger scientists. He had a delightful sense of humour, and in personality was honourable and unassuming, though somewhat prone to irritation with those whose views on the quantum theory were not as pure as his own. He died suddenly of a stroke in October 1990.

Further Reading

Speakable and Unspeakable in Quantum Mechanics, J S Bell (Cambridge University Press, 1987) contains most of Bell's quantum papers.

Quantum Mechanics, High Energy Physics and Accelerators, J S Bell (World Scientific, Singapore, 1995) (edited by M Bell, K Gottfried and M Veltmann) contains many papers by Bell on all these areas.

Quantum Profiles, J Bernstein (Princeton University Press, 1991) contains substantial accounts of interviews with John Bell.

Einstein, Bohr and the Quantum Dilemma, A Whitaker (Cambridge University Press, 1996) and 'Theory and Experiment in the Foundations of Quantum Theory', A Whitaker, *Progress in Quantum Electronics* **24** (2000) 1–106, give general accounts of the area.

Lochlainn Ó Raifeartaigh

1933–2000

Siddhartha Sen

Lochlainn Ó Raifeartaigh was an important figure in fundamental theoretical physics. He was widely respected for his many contributions to high energy physics, where he often used ideas of symmetry and group theory to great effect. The list of his published work contains over 200 entries, many in collaboration. The list of his collaborators is over 60. Ó Raifeartaigh is best known for his no-go theorem of 1965 and his discovery of a way to spontaneously break supersymmetry in 1975.

To explain these results and their significance a few preliminary remarks about fundamental theoretical physics are necessary. One of the goals of fundamental physics is to discover the laws that govern the multitude of observed phenomena. The process for establishing a law is thorough. The law proposed is first used to deduce experimentally testable consequences. These consequences are then looked for. If found the law becomes more plausible. Repeated successful tests eventually establish the proposed law as a very useful way of encoding information about nature.

This process of unravelling the secrets of nature has step by step forced us to change profoundly the way we think of space, time and matter. Progress has often been slow and has depended on ingenious experiments, the interpretation of which by the giants of science such as Galileo, Kepler, Newton, Faraday and Maxwell has gradually led to a coherent picture of the workings of nature.

By the end of the nineteenth century and in the early twentieth century it was established that matter was made out of atoms and that atoms were in some sense like small planetary systems consisting of a positively charged nucleus around which the negatively charged electrons revolve endlessly. Atoms emitted light. It soon became clear that neither the stability of the atom nor the details of the way the atom emitted light could be understood using the ideas of Newton regarding motion and of Maxwell regarding the way charged particles behave when accelerated.

To explain why this was the case let us consider the simplest of atoms, namely hydrogen. Hydrogen has a central positively charged

Lochlainn Ó Raifeartaigh (courtesy of Treasa Ó Raifeartaigh).

proton with a single negatively charged electron circling around it. A circling electron according to Newton was in a state of acceleration. This implied, according to Maxwell, that the particle must lose energy by emitting light. As the electron lost its energy it would spiral down continuously to the central proton. The time taken for the process of collapse could be calculated and shown to take only a fraction of a second. The planetary model of hydrogen is thus not stable. The final state of the atom predicted by the ideas of Newton and Maxwell is of the electron and proton with zero distance between them, thus forming an atom the size of a proton. Experimentally such a model was ruled out by Rutherford. It was also the case that the pattern of light emitted by hydrogen could not be understood.

The resolution of these problems came from the revolution of quantum theory. Quantum theory changed the way nature was described. Classical certainty present in the theories of Newton and Maxwell was replaced by quantum uncertainty, determinism was replaced by probability, local realism was replaced by non-local effects which were contrary to everyday intuition.

These insights from quantum theory also profoundly modified the idea of the classical trajectories that particles were expected to follow and showed how the uncertainty principle of quantum theory, suitably generalized, could explain why the electron could no longer spiral down to the central proton and thus account for the stability of the hydrogen atom. The pattern of light emitted by hydrogen could also now be understood.

The quest for understanding the structure of matter led to a study of the atom and then, as experimental techniques improved, progressed to the study of the structures present deep inside the nucleus of the atom. Progress in our understanding of the nature of matter in recent times has been spectacular. A theory providing a unified picture of the way the electromagnetic, the weak and the strong forces all fit together known as the standard model has emerged. The electromagnetic forces are the long range forces responsible for holding atoms and molecules together, the weak forces are short range forces responsible for radio-activity while the strong forces are short range forces responsible for holding together the quarks which are the constituents of nucleons. Using this theory the world inside the atom is now understood. It is governed by the laws of quantum theory.

The analysis of space and time in the twentieth century also led to surprising conclusions. Instead of the certainty of *a priori* Euclidean geometry, the work of Einstein suggests a picture in which the fabric of space and time was itself to be determined dynamically. Instead of being a backdrop for events, space–time was henceforth to be regarded as a participant in the drama of physics. Understanding space–time in this sense meant understanding the force of gravitation in geometrical terms since the change in the fabric of space–time in the presence of matter manifests itself as the gravitational force.

These two directions of inquiry, the nature of space–time (the theory of gravitation) and the nature of matter (the study of the weak, electro-magnetic and strong forces) were pursued separately by the majority of the scientific community. It was Einstein's dream to forge a synthesis. In Einstein's vision, the quantum idea, essential for understanding the properties of atoms, was regarded as an incomplete but useful description of nature. Eventually, Einstein felt, the disturbing non-local features of quantum theory would not be present in the final unified theory. As yet no viable candidate theory of this kind has been proposed.

In view of this, the alternative approach of trying to see if the theory of gravitation can be accommodated in a quantum framework needs to be considered. Indeed in recent times encouraging progress in constructing such a quantum theory of gravitation has been made. A promising quantum theory of space–time and matter known as string theory is now being actively pursued worldwide. A key idea for the consistency of string theory is that it should have a novel symmetry known as super-symmetry. Supersymmetry is a profound quantum symmetry which brings together the two basic classes of observed fundamental particles. One class consists of particles known as fermions (named after the Italian physicist Enrico Fermi). These are the particles which constitute matter. In the standard model these are the quarks which are the building blocks of the proton, neutron and the other strongly interacting particles and the leptons which interact weakly or electomagnetically. The second class of particles consists of gauge bosons (named after the Indian physicist Satyen Bose). These are the particles responsible for forces. Thus in the standard model the gauge bosons that hold the quarks together are known as gluons (they glue quarks together). The gauge boson that gives rise to electromagnetic forces is the photon, while the gauge bosons responsible for the weak forces are known as the W and Z boson. Supersymmetry requires that each boson must have a supersymmetric fermionic partner and each fermion must have a supersymmetric bosonic partner. The supersymmetric partner of the electron is a boson called the selectron and the supersymmetric partner of a gauge boson is a fermion called a gaugino. Supersymmetry has been shown to have deep connections with ideas in mathematics. However there is as yet no direct experimental evidence for supersymmetry.

The theoretical and indirect reasons for regarding supersymmetry as a very important fundamental symmetry are, however, compelling. The world as we see it is certainly not supersymmetric; thus if supersymmetry is indeed a fundamental symmetry of nature it can only be a 'hidden symmetry' which is broken in the observed world. Even though the symmetry is broken, the fact that it is present at a fundamental level has far reaching consequences. It is, for instance, essential for the consistency of string theory. The power, novelty and fundamental nature of supersymmetry was recognized as soon as it was discovered by Wess and Zumino in 1974. One of its surprising aspects was that it profoundly changed the nature of space. Space was replaced by superspace.

The hunt to find a way to incorporate this new symmetry as a hidden symmetry immediately began. Ó Raifeartaigh solved this problem in 1975. The mechanism of how a spontaneously broken hidden symmetry could occur in a relativistic theory was clearly explained by Peter Higgs in 1964. In a spontaneously broken symmetry situation the underlying theory has the symmetry but the physical system considered ('the ground state') does not.

Higgs was examining the way forces between massless fermions could be introduced by using a symmetry principle known as the gauge principle. The gauge principle was originally discovered in electromagnetism and provided a beautiful geometric reason for the presence of the electromagnetic forces. It was thus natural to ask if the weak and the strong forces could be understood by suitably extending the gauge principle.

The simplest attempts to do this ran into a problem. The forces introduced by the gauge principle all seemed to be long range forces.

Experimentally it was well established that the weak and strong forces were short ranged. Hence the beautiful geometric gauge principle did not seem to be capable of providing the unifying theme necessary for understanding the strong and the weak forces. This widely regarded viewpoint was modified by the work of Higgs. Higgs showed in his theoretical model that the gauge principle in the presence of a new boson (the Higgs boson) introduced with a suitable self-interacting potential (the Higgs potential) had a remarkable property. The system could spontaneously break its gauge symmetry and in the process make the fermions of the theory massive (even though originally the fermions were required to be massless by the gauge principle) and at the same time introduce forces which were short ranged.

These amazing possibilities were later used by Weinberg, Salam and Glashow to construct a theory unifying the weak and electromagnetic forces using a gauge principle and a generalization of Higgs' ideas. This work predicted the existence of two new bosons, the W and the Z, which were subsequently discovered. Weinberg, Salam and Glashow were awarded the Nobel prize for their work.

The problem Ó Raifeartaigh tackled was to see if a generalization of Higgs' ideas was possible to spontaneously break supersymmetry. The feeling in the scientific community at that time was such a generalization might not be possible. This was based on the failure of attempts to generalize Higgs' ideas to supersymmetry in the simplest possible way. Ó Raifeartaigh carried out a careful analysis of different generalizations of Higgs' ideas and found a class of ways to spontaneously break supersymmetry. This work immediately allowed theorists to contemplate supersymmetry as the underlying grand symmetry of nature. A major hurdle had been overcome.

Supersymmetric theories had another very special feature. They were able to combine internal symmetries with the space–time symmetries of special relativity in an interesting way. A simple example of a space–time symmetry is rotational invariance. If gravitational effects are ignored then different directions of space are equivalent. This observation has a theoretical implication. It introduces patterns in the masses which the fundamental particles can have that are found to be true experimentally. Internal symmetries extend these ideas in a profound way. To explain what is involved let us consider two strongly interacting particles, namely the neutron and the proton. These particles are the constituents of nuclei. If electric forces are ignored, the proton and neutron have very similar properties. It was suggested by Heisenberg that if a two-dimensional internal space called isospin was introduced, one direction of which represented a proton while the other direction represented a neutron, and rotational invariance in this internal space was postulated, then patterns in the way the neutrons and protons behave could be predicted. These predicted patterns were found to be in agreement with experimental observations.

In a supersymmetric model these two symmetries, namely space–time symmetries of special relativity theory and internal (isospin-like) symmetries, could seamlessly be joined together. This was an important result. Using conventional symmetries known as Lie group symmetry, it had been shown theoretically that this was not possible. Indeed this was the content of the no-go theorem established by Ó Raifeartaigh in 1965.

Prior to 1961 theoretical physicists used symmetry methods mainly to classify crystalline structures revealed by X-ray studies. Symmetry methods were also used to understand the general pattern of light that was emitted by atoms, molecules and nuclei. The underlying symmetry used here was that of a system under rotations.

In 1961 the situation changed utterly. This revolution in the use of symmetry came from Murray Gell-Mann. Gell-Mann introduced a new symmetry which went under the name SU(3). The symmetry was introduced to bring order to the enormous number of 'fundamental' particles that were being experimentally discovered at that time. Using SU(3), Gell-Mann was able to group particles together in well defined patterns. In one of the patterns there was a missing slot. This suggested to Gell-Mann that a new particle called the Ω^- had to be present if SU(3) was correct. Gell-Mann was able to predict the exact mass, the charge and the spin of the Ω^- particle from his pattern. The Ω^- particle was discovered in 1962 soon after Gell-Mann's prediction.

This made the theoretical physics community rush to learn group theory and see if other patterns could be discovered. The standard source for learning the mathematics necessary at that time was a set of lecture notes given by Racah at Princeton in 1961. What Gell-Mann had discovered was an extension of the internal isospin symmetry of Heisenberg. Isospin was based on an analogy of rotations in real space with rotations in an internal isospin space. Gell-Mann's work extended the idea to a new symmetry completely different from the symmetry of rotations.

Soon extensions of Gell-Mann's idea started to appear. The extensions were of two kinds. Symmetries other than SU(3) were considered and attempts to combine Gell-Mann's internal symmetry with that of space–time, fusing the two together in an interesting way, amenable to experimental verification. Such a promising combined symmetry was soon proposed involving the group SU(6). In this symmetry, the internal symmetry of Gell-Mann was successfully fused with rotational symmetry of space. Now the hunt to find the best possible extension of Gell-Mann's idea heated up. The prize was to discover a symmetry that combined the internal symmetry of Gell-Mann with the full Poincaré symmetry of space–time associated with Einstein's special theory of relativity, as it was well known that rotational symmetry of space was only a part of Poincaré symmetry. In the midst of furious attempts to crack this prize problem, Ó Raifeartaigh showed that under very general conditions the problem posed had no useful solution. The methods used by Ó Raifeartaigh to prove his result were subtle and made use of deep results from the theory of Lie groups way beyond the topics covered in Racah's Princeton lectures and hence unfamiliar to most physicists. This work brought to an abrupt end major efforts to combine internal and Poincaré symmetries.

This 'no-go' theorem also established Ó Raifeartaigh as an expert in symmetry theory and as a researcher of the highest rank. Recalling the events surrounding Ó Raifeartaigh's work, McGlinn, who was the first to formulate the symmetry problem in group theoretic terms and the first to point out the difficulty of joining together internal and Poincaré symmetry, said "I felt Lochlainn was going to kill the program. He was sharp and his knowledge of group theory was way ahead of anything I or most other physicists knew at that time."

Ó Raifeartaigh was at Syracuse University, USA, when he proved his no-go theorem. A few years later he returned to Dublin as a Senior Professor of theoretical physics at the Dublin Institute for Advanced Studies (DIAS) where he spent the rest of his life working on fundamental problems, arguing and discussing physics with passion, training post-doctoral fellows and inspiring postgraduate students by his lectures on gauge theory and group theory.

Lochlainn Ó Raifeartaigh was born in Clontarf, Dublin, on 11 March 1933. He was the second eldest of a family of three brothers and three sisters. His father Tarlach Ó Raifeartaigh was a Gaelic scholar and civil servant who as Secretary of the Department of Education and first Chairman of the Higher Education Authority was instrumental in the development of university education in Ireland. His mother Neans was a daughter of T J Morrissey, a first class mathematics graduate from the Royal University who went on to become the Assistant Secretary of the Department of Education. She spoke Irish and French fluently and had excellent knowledge of German, Swedish and Japanese.

Ó Raifeartaigh attended St Joseph's primary school in Marino and then attended Castleknock College where he enjoyed the atmosphere and the beautiful setting, and participated enthusiastically in all sporting activities available. In 1950 he completed school, securing first place in Mathematics in the Irish Leaving Certificate Examination and winning a University Entrance Scholarship to University College Dublin (UCD). He had decided to study science, abandoning his original intention to study engineering. Writing in the *Castleknock Chronicle* Ó Raifeartaigh recalled that "as 1950 was a Holy Year I and four classmates conceived the mad idea of spending that summer holiday cycling to Rome". The love of the outdoors and adventure this reveals was to remain with him for the rest of his life.

One of his interests while a student at UCD had an important impact on his life. This was his interest in the Irish language which led him to join An Cumann Gaelach (the Gaelic Society). Members of An Cumann Gaelach from all the Irish Universities, north and south, met once a year in different parts of the Gaeltacht (native Irish-speaking districts). Lochlainn went to Teelin near Donegal, where he met Treasa Donnelly, a student from Queen's University Belfast, whom he married in 1958. Treasa had a love of learning languages and shared Lochlainn's interest in the theatre and hill walking. She provided the support which allowed him to concentrate his energies fully on scientific research. The five children Conor, Finbar, Cormac, Una and Aoife were a source of pride and joy to him and to Treasa.

After graduating from UCD in 1956, Ó Raifeartaigh obtained a research fellowship from DIAS where he studied with the great Irish relativist Synge*. The Dublin Institute for Advanced Studies was created by the Taoiseach (Prime Minister) of Ireland Éamon de Valera as an Institute where fundamental research at the frontiers of knowledge was to be carried out and where young researchers would be trained in the methods of fundamental research. De Valera was able to persuade Nobel Prize winner Schrödinger* to come to Dublin as the first Director of DIAS. A galaxy of talent flowed to Dublin. Heitler*, Synge, Lanczos* and Takahashi were members of the faculty. After the war some moved back to their home countries. Heitler left DIAS for Zurich in 1949 and Schrödinger for Vienna in 1956.

After a year at DIAS and three publications in relativity theory Ó Raifeartaigh was awarded a travelling studentship to study under Heitler in Zurich. The University of Zurich had a distinguished record of excellence in theoretical physics. It was after all the place where Schrödinger discovered quantum mechanics! After completing his PhD in 1960 Ó Raifeartaigh returned to DIAS as an Assistant Professor in 1961.

A turning point in Ó Raifeartaigh's life came as a result of a chance encounter with E C G Sudarshan at Berne. As a direct result of this meeting Ó Raifeartaigh was invited to spend three months at the newly formed Mathematical Science Institute in Madras to lecture on group theory in 1963. The rapport with Sudarshan led to Ó Raifeartaigh being invited by Sudarshan to join his group at Syracuse, USA, in 1964 on extended leave from DIAS. It is here that Ó Raifeartaigh used his considerable knowledge of group theory to prove his no-go theorem. The chance encounter at Berne had had important consequences!

Even though Ó Raifeartaigh and his family enjoyed living at Syracuse, they were keen to return to Europe. When the opportunity to return to Dublin as Senior Professor at DIAS came up in 1968, Ó Raifeartaigh readily accepted. Here he built up an internationally respected group in theoretical physics working at the frontiers of the subject. As new concepts emerged in theoretical particle physics Ó Raifeartaigh made them his own and provided important contributions to them.

His method of training postdoctoral fellows was to meet with them regularly and to interact with them in front of a blackboard. The discussions were not polite considered exchanges but heated far-ranging arguments which continued until the issues being debated had been clarified. The aim of these discussions was to understand the essence of the problem being considered. Ó Raifeartaigh often said that the main difficulty in tackling a problem was to properly understand it. These far-ranging, thorough, open-ended discussions are fondly remembered by those who participated in them.

Ó Raifeartaigh had wide interests and was fluent in Irish, French and German. He was a scholar in the history of his subject and wrote a masterly book *The Dawning of Gauge Theories* in which the historical quest for a unified theory of the fundamental forces of the physical world is described. As he says in the preface: "It is hoped that the book will be of value not only as a description of how modern gauge theory developed in its early days but also in showing how original ideas develop and come to fruition in spite of difficulties and frustrations."

Ó Raifeartaigh received a von Humboldt Research Award in 1998 and the prestigious Wigner Medal in 2000 for his pioneering contributions to particle physics. He was elected a Member of the Royal Irish Academy at the age of 29, and was also a member of Academia Europaea. His gift of clear exposition led to frequent invitations to lecture at summer schools and conferences. Often he was accompanied by his family on longer visits. Two delightful sabbaticals were spent with his family at the Institut des Hautes Études Scientifiques at Bures-sur-Yvette, France. Ó Raifeartaigh's international stature also led to invitations to spend extended periods of time in many countries including Sweden, Germany, Italy, France, Switzerland, USA and twice in recent years in Japan. He was also interested in the international aspects of science and was a conscientious founder member of the Irish Pugwash group.

Ó Raifeartaigh was an avid theatregoer, enjoyed music and was a dedicated member of a local ramblers club Na Coisithe. No matter what the weather he would be out walking every second Sunday accompanied by Treasa and friends. It is perhaps appropriate to close our brief account of the life of this quiet, modest, profound scholar, Irish by birth but international in his outlook, by quoting from the notice of his death that appeared in his ramblers club newsletter: "On November 18, 2000

Ó Raifeartaigh speaking at the conference 'Group Theoretical Methods in Physics' held in Dubna near Moscow in 2000, at which he was awarded the Eugene Wigner Medal (courtesy of Treasa Ó Raifeartaigh).

Na Coisithe suffered the sad unexpected loss of our friend Lochlainn Ó Raifeartaigh. A popular but most unassuming man, few of his walking friends were aware that Lochlainn was a physicist with an international reputation of the highest calibre.''

Acknowledgment
I am grateful to Mrs Treasa Ó Raifeartaigh, Mr John Gardiner and Professor William McGlinn for their help. Mrs Ó Raifeartaigh also kindly helped with illustrations.

Further Reading

The Quantum Universe, Tony Hey and Patrick Walters (Cambridge University Press, 1987).
Dreams of a Final Theory: The Search for the Fundamental Laws of Nature, Steven Weinberg (Vintage, London, 1993).
The Supersymmetric World, G Kane and M Shifman (World Scientific, Singapore, 2000).

Index of Names

With the exception of a few well known figures from history and letters, only names of scientists and some scientific institutions are included. Page numbers of articles are indicated in bold type; pages on which pictures occur are marked with an asterisk.